Pietro Snider
The Natural Problem of Consciousness

Epistemic Studies

Philosophy of Science, Cognition and Mind

Edited by
Michael Esfeld, Stephan Hartmann, Albert Newen

Volume 36

Pietro Snider

The Natural Problem of Consciousness

DE GRUYTER

ISBN 978-3-11-065367-0
e-ISBN (PDF) 978-3-11-052557-1
e-ISBN (EPUB) 978-3-11-052469-7
ISSN 2512-5168

Library of Congress Cataloging-in-Publication Data
A CIP catalog record for this book has been applied for at the Library of Congress.

Bibliographic information published by the Deutsche Nationalbibliothek
The Deutsche Nationalbibliothek lists this publication in the Deutsche Nationalbibliografie;
detailed bibliographic data are available on the Internet at http://dnb.dnb.de.

© 2019 Walter de Gruyter GmbH, Berlin/Boston
This volume is text- and page-identical with the hardback published in 2017.
Printing and binding: CPI books GmbH, Leck
♾ Printed on acid-free paper
Printed in Germany

www.degruyter.com

Acknowledgments

First and foremost I thank my dissertation supervisors Prof. Markus Wild and Prof. Michael-Andreas Esfeld for trusting me, for their relentless, competent, and frank professional support and advising throughout the years, for their patience in accompanying me through administrative endeavours, and most of all for their friendliness and humanity. This work would have not been possible without their constant encouragement. I thank the Swiss National Science Foundation for generously funding my research in Switzerland and abroad, allowing me to pursue my interests in very privileged working conditions. I also thank the Forschungsfonds der Universität Basel for allowing me to conclude my dissertation in the best possible conditions.

I owe a lot to the students, faculty, and staff of the philosophy departments of the Université de Lausanne, Université de Fribourg, and Universität Basel. For providing a pleasant and intellectually stimulating working environment over the years I wish to mention in particular Emmanuel Baierlé, Jiry Benovsky, Simone Chambers, Christine Clavien, Coralie Dorsaz, Matthias Egg, Patrik Engisch, Andrea Giananti, Nadja Heller Higy, Brigitte Hilmer, Gunnar Hindrichs, Rebekka Hufendiek, Thomas Jacobi, David Jolidon, Susanne Kress, Vincent Lam, Anna Lettieri-Beck, Hannes Ole Matthiesen, Anne Meylan, Deborah Mühlebach, Franziska Müller, Jan Müller, Jacob Naïto, Martine Nida-Rümelin, Michael O'Leary, Sebastian Pabst, Matthieu Queloz, Christian Sachse, Melanie Sarzano, Mario Schärli, Susanne Schmetkamp, Jelscha Schmid, Hubert Schnüringer, Christine Sievers, Gianfranco Soldati, Michael Sollberger, Marc Nicolas Sommer, Patrice Soom, Christian Steiner, Marie van Loon, and Antonio Vassallo.

I am very grateful to Albert Newen, Karen Neander, and Ned Block for supervising me during my research visits, for their constructive criticism, the priceless commentaries, the suggestions for improvement, and – more in general – for offering me the great opportunity to improve my work in highly stimulating environments. I am also indebted to the rest of the faculty, the graduate students and the staff at the Institut für Philosophie II and Center for Mind, Brain and Cognitive Evolution at the Ruhr-Universität Bochum, the Philosophy Department of Duke University, and the Philosophy Department of NYU. My research visits would not have been equally enriching, profitable and rewarding both at a professional and personal level without the company and the discussions I had with David Barack, Luca Barlassina, Rosa Cao, Grace Helton, Lena Kästner, Andrew Lee, Vincent Legeay, Kristen Miller, Jorge Morales, Nicolas Porot, Tomoo Ueda, Petra Vetter, and many others. I also would like to thank the audiences of talks on intermediate versions of this work I gave in Fribourg, Luzern, Coglio, Mainz, Bochum, Ovronnaz, Geneva, Lausanne, San Diego, Basel, Durham (NC), New York, and

Tucson for helping me improve my work with precious comments, questions and suggestions. These include also some of the experts I had the pleasure to talk to during conferences, workshops, and in private correspondence during the last years, including Fred Adams, David Chalmers, Tim Crane, Daniel Dennett, Owen Flanagan, Peter Godfrey-Smith, Steven Harnad, Robert Hopkins, Frank Jackson, Thomas Metzinger, Thomas Nagel, Jesse Prinz, Jim Pryor, Chris Peacocke, David Rosenthal, John Searle, Marcel Weber, and Michael Wheeler.

Many thanks to Paolo Jacomelli for his meticulous proofreading, to Konrad Vorderobermeier for his support and care in the formatting of the manuscript, to Stephan Hartmann, Albert Newen and Michael Esfeld, editors of the series Epistemic Studies, and to De Gruyter – in particular Christoph Schirmer, Gertrud Grünkorn, Nancy Christ, and Maik Bierwirth – for their kind and professional support throughout the process leading to the publication.

Last, but not least, I wholeheartedly thank my family and friends for their unconditional love and invaluable support throughout all these years. This work is dedicated to all of you.

Pietro Snider
Locarno, 8 March 2017

Funding

This work has been supported by the Swiss National Science Foundation and the Forschungsfonds der Universität Basel.

Junior Research Grant, Forschungsfonds der Universität Basel, Förderung exzellenter junger Forschender 2015 (no. DGK2615).

SNSF Mobility Grant (Mobility Within Project no. 139037) for a research stay at NYU and Duke University, including the participation to the Tucson Consciousness Conference 2014, United States (03.01.2014 – 30.06.2014).

University of Fribourg Grant for participation to the 17th Conference of the Association for the Scientific Study of Consciousness (ASSC), San Diego, USA (July 2013)

SNSF Project no. 139037, "Biosemantics and Normative Pragmatism: Towards a Unified Picture of the Place of the Mind in the Natural World" (Professeurs boursiers FNS). Main applicant Prof. Markus Wild. Université de Fribourg and Universität Basel (15.07.2012 – 30.09.2014).

University of Lausanne Grant for participation to the Summer School in the Cognitive Sciences "Evolution and Function of Consciousness", Montréal, Canada (June–July 2012)

SNSF Project no. 132389, "Causal Properties and Laws in the Philosophy of Science" ("Mind and Reality" ProDoc). Main applicant Prof. Michael Esfeld. Université de Lausanne (01.10.2010 – 30.06.2012).

Contents

Acknowledgments —— V

1 Introduction —— 1
    1.1 Why Care About Consciousness —— 1
        1.1.1 Nothing To Care About —— 3
        1.1.2 An Intellectual Challenge —— 4
        1.1.3 Practical and Ethical Applications —— 4
        1.1.4 A Hopeful Research Project in Philosophy and Science —— 5
    1.2 Consciousness and the Scientific Paradigm —— 8
        1.2.1 Epistemic Objectivity and Ontological Subjectivity —— 9
        1.2.2 Dealing With the Subjective Ontology of Consciousness —— 13
    1.3 Neural Correlates of Consciousness (NCCs) —— 16
        1.3.1 Defining NCC —— 18
        1.3.2 Explanatory Limits and Fuzziness of NCC Research —— 21
    1.4 The Hard Problem of Consciousness —— 24
    1.5 Consciousness: Toward A Diachronic Approach —— 28
        1.5.1 Consciousness in Time: a Thought Experiment —— 31
        1.5.2 Excursus on Objective Knowledge —— 34

2 The Metaphysical Problem of Consciousness —— 40
    2.1 The Mind-Body Problem: the Mind in a Physical World —— 40
        2.1.1 The Set of Premises to the Puzzle of the Inconsistent Triad —— 41
        2.1.2 Only a Physicalist Framework Can Account for Causation —— 42
        2.1.3 The Subjective Character of Consciousness is Irreducible —— 43
        2.1.4 Mental Causation —— 50
        2.1.5 The Joint Inconsistency of the Set of Premises —— 51
    2.2 A Sketch of the Metaphysical Landscape —— 53
        2.2.1 Ontological Monism —— 54
        2.2.2 Ontological Dualism —— 58
        2.2.3 Supervenience as Framework for Physicalism —— 61
    2.3 The Metaphysical Background for the Natural Problem —— 64

3 The Natural Problem of Consciousness —— 66
    3.1 Adopting Naturalism: A Pragmatic Approach —— 67

	3.2	Consciousness as Contingent Biological Phenomenon —— 70	
		3.2.1 Dismissing Solipsism in the Actual World —— 73	
		3.2.2 Dismissing Panpsychism in the Actual World —— 74	
	3.3	Why Are There Feeling Beings in the Natural World? —— 76	

4 Consciousness as Feeling. Defining Criteria —— 80
- 4.1 State, Creature, and Being Consciousness —— 80
  - 4.1.1 State Consciousness —— 82
  - 4.1.2 Creature Consciousness —— 86
  - 4.1.3 Being Consciousness —— 88
  - 4.1.4 Intentionality and Transitivity —— 88
  - 4.1.5 Higher-Order Theories of Consciousness —— 90
- 4.2 Attempts to Characterize Consciousness —— 92
  - 4.2.1 Consciousness: A Mongrel Concept —— 92
  - 4.2.2 Toward a Taxonomy of Intransitive Being Consciousness —— 95
  - 4.2.3 Candidate Feeling Beings: A Problem —— 97
- 4.3 Defining Criteria for Being Consciousness —— 98
  - 4.3.1 Limits of Characterizations of Consciousness —— 98
  - 4.3.2 Consciousness as Sentience and Responsiveness —— 101
  - 4.3.3 Consciousness as Wakefulness and Normal Alertness —— 103
  - 4.3.4 Phenomenal Consciousness (What it is Like) —— 106
- 4.4 Feeling and Self-Consciousness —— 110

5 Working Out Diachronic Claims —— 115
- 5.1 Feeling Presently Exists —— 115
- 5.2 Phylogenetic Evolution and Ontogenetic Development —— 120
  - 5.2.1 Phylogenetic Evolution —— 120
  - 5.2.2 Ontogenetic Development —— 121
- 5.3 Two Fine-Grained Questions —— 122
  - 5.3.1 Radical Change Claim VS Qualitative Change Claim —— 124
- 5.4 Examining the Two Fine-grained Diachronic Claims —— 126
  - 5.4.1 Feeling Is Subject to Phylogenetic Evolution (a) —— 126
  - 5.4.2 Feeling Is Subject to Ontogenetic Development (b) —— 131
- 5.5 Summary of the Claims and of the Conclusions —— 136

6 Why Do We Feel? —— 137
- 6.1 Feeling in an Evolutionary Framework —— 137
  - 6.1.1 A Darwinian Evolutionary Theory —— 137
  - 6.1.2 A Sketch of the Hypothesis —— 143
- 6.2 Distinguishing Kinds of Function —— 144

- 6.3 The Evolutionary Function of Feeling —— 146
  - 6.3.1 Feeling as Nonadaptation —— 148
  - 6.3.2 Feeling as Maladaptive Trait —— 148
  - 6.3.3 Feeling as Adaptation —— 148
  - 6.3.4 Feeling as Exaptation —— 149
  - 6.3.5 Feeling as Preadaptation —— 149
  - 6.3.6 Feeling as Spandrel —— 149
  - 6.3.7 Conclusion on the Evolutionary Function of Feeling —— 150
- 6.4 Could Feeling Have a Biological (Causal) Function? —— 151
  - 6.4.1 Epiphenomenalism —— 151
  - 6.4.2 Reasons To Put Epiphenomenalism Aside —— 153

7 A Hypothetical Biological Function of Feeling —— 155
- 7.1 How To Settle the Question: A Methodological Point —— 155
- 7.2 Bottom-up Approach —— 157
- 7.3 Biological Function: a Top-Down Hypothesis —— 165
  - 7.3.1 Biological Efficacy and Strategies to Maximise Profit —— 165
  - 7.3.2 Global *Qualia* and Multiple Realization —— 167
  - 7.3.3 Similarities Across Global *Qualia* —— 171
  - 7.3.4 The Example of Valence —— 173
  - 7.3.5 The Reliability Of Valence Feelings —— 177
  - 7.3.6 Valence Feelings and Causation —— 182
  - 7.3.7 Learning to Behave Effectively —— 184
  - 7.3.8 Summary of the Hypothesis and Preliminary Conclusion —— 189

8 Causation and the Conscious Mind —— 191
- 8.1 Theories of Causation —— 191
  - 8.1.1 Causation as Constant Conjunction —— 192
  - 8.1.2 Causation as Counterfactual Dependence —— 196
  - 8.1.3 Causation as Transfer of Physical Quantities —— 198
  - 8.1.4 Viable Theories of Causation —— 199
- 8.2 Testing the Consciousness-Behaviour Causal Relation —— 199
- 8.3 A Conservative Conclusion to the Natural Problem —— 204

Appendix: Objections and Replies —— 207

Bibliography —— 223

Name Index —— 235

1 Introduction

The main goal of the present dissertation is to introduce and unpack the following question: Why are there presently conscious beings at all? How can we explain the fact that humans and other animals have turned out to be conscious beings, rather than evolving as physical systems with no conscious mental life?

In order to best prepare the ground for the detailed articulation, development, and assessment of my contribution in chapter 3 and following, I devote the first two chapters to contextualising the setup of the discussion. In chapter 1 I introduce some important general issues concerning the nature, study and ways of thinking about problems of consciousness, whereas in chapter 2 I introduce the metaphysical mind-body problem and the working assumptions I endorse in the rest of the dissertation.

I begin by explaining why I think it is worth studying and asking philosophical questions about consciousness in the first place (section 1.1). Then, I argue that science can deal with consciousness (section 1.2). I present the quest for Neural Correlates of Consciousness and discuss some of its limits (section 1.3). I dedicate special attention to the "Hard Problem" of consciousness, clarifying that I am not concerned with that problem (section 1.4). I conclude by claiming that the above-mentioned approaches and problems of consciousness rise within a synchronic paradigm, and that the research question I intend to tackle stems from an alternative, often-neglected, diachronic way of thinking about consciousness (section 1.5).

1.1 Why Care About Consciousness

> *Perhaps no aspect of mind is more familiar or more puzzling than consciousness and our conscious experience of self and world.* (Van Gulick 2014)

Consciousness is an essential feature of the human mind. Even though the specific conceptions of what consciousness is vary – sometimes dramatically –, an impressive array of philosophers and psychologists has independently dealt with it one way or another, suggesting that besides the disagreements about what consciousness is, or how we should study it, there is a widespread tacit agreement that consciousness ought to be studied and accounted for. Notable western scholars that dealt with consciousness from the seventeenth century until the beginning of the twentieth century include philosophers such as Descartes (1644), Locke (1688), Leibniz (1686; 1714), Hume (1739), Kant (1787), James Mill (1829), John Stuart Mill (1865), Husserl (1913), Heidegger (1927), and Merleau-Ponty

(1945), as well as modern scientific psychologists such as Wundt (1897), von Helmholtz (1910), James (1890) and Titchener (1901)[1]. Despite a quiet period for the study of consciousness due first to the rise of behaviourism in scientific psychology (Watson 1924; Skinner 1953), and then to the rise of cognitive psychology (Neisser 1965; Gardner 1985), since the 1980s there has been a major resurgence of philosophical and scientific research into the nature of consciousness (Lycan 1987, 1996; Baars 1988; Penrose 1989, 1994; Dennett 1991; Crick 1994; Chalmers 1996). The fact that during the past thirty-five years consciousness has become a more and more popular and widely accepted subject of philosophical and scientific research, and that today there is an across-the-board interest on this subject matter ranging from armchair philosophy, to psychology and neuroscience, does not only indicate that consciousness is again a fashionable subject of research. It also suggests, more profoundly, that tackling this subject matter is somehow crucial to fully understanding human nature, and that the intellectual enterprise of understanding the relationship between mind, brain and natural world requires accounting for consciousness one way or another.

Even if the historical treatment of this subject matter suggests that consciousness plays an important role in the interplay between mind, brain and natural world, it might be argued that this widespread and long-lasting interest in consciousness does not by itself clarify what exactly is the *motivation* for taking consciousness-related issues seriously and for investigating consciousness further. After all, it might be argued, the importance of consciousness might simply have been overstated without any good reasons, and perhaps the way out of the problem of consciousness consists simply in not worrying about it. The reluctance to concede that consciousness is a valuable object of interest and research could be partially a manifestation of skepticism towards philosophical research in general – linked to the idea that studying consciousness can only be of interest to philosophers –, and partially a consequence of the bewilderment and lack of clarity surrounding this particular topic. I think that the objection is not pertinent, the reluctance unjustified, and that if we aim to achieve a satisfactory scientific understanding of the world we have to take consciousness seriously. However, in order to preventively counter any off-putting attitude or skepticism threatening to diminish the perceived value of any serious research on consciousness, it is worth considering the objection seriously giving reasons explaining *why* the importance of consciousness is not an overstatement. Appreciating the motivations for studying consciousness should be enough to get past these initial hitches and will serve both as motivation vindicating my decision to work on con-

---

1 I draw from the historical account in Van Gulick (2014, section 1).

sciousness in the first place and as a preliminary survey of some of the issues at stake.

What reasons are there to keep on tackling the captivating but slippery walls of a controversial and hotly debated topic such as consciousness? Why would it make sense to keep on working on better descriptions, a better understanding, and better explanations of consciousness? I think there are four main reasons. *First*, consciousness underlies every aspect of our life that we take as having a quality or meaning, i.e., every conscious experience[2]. Since experience is that in virtue of which we value and treasure life, there is a good *prima facie* reason to study consciousness (section 1.1.1). *Second*, studying consciousness is complex and challenging. This makes it a stimulating intellectual enterprise, since there is much to do, and virtually none of it is dull research work (section 1.1.2). *Third*, studying consciousness can lead to both theoretical and practical gains. The possibility of achieving knowledge about a fundamental aspect of our life while also having an impact on how to deal with consciousness-related matters in every day life is a strong motivation to work on consciousness (section 1.1.3). *Fourth*, it is possible to study consciousness scientifically, and we are getting better at that. The science of consciousness has progressed hugely and constantly in the past few decades, and there are reasons to be optimistic about the future potential of scientific consciousness research (section 1.1.4). These four reasons taken together suggest that it is worth caring about consciousness and taking it seriously as object of study and research, and that now more than ever before there are good odds of making serious progress in the understanding of consciousness.

1.1.1 Nothing To Care About

A first reason to take consciousness seriously is that in the actual natural world consciousness[3] is contingently linked to the mattering of things, of any thing. Anything that matters to us – whether our beloved ones, health, scientific

---

[2] I am sympathetic towards the common sense conception of the term "experience" according to which experience is always conscious (cf. Strawson 1994). Thus, when I talk about "experience" *simpliciter*, this can be read as "*conscious* experience". Since there is a different conception according to which experience is not always conscious (cf. Carruthers 1989, 2000), in order to avoid confusion, if I want to refer to cases in which experience is not conscious I explicitly call it *unconscious* or *non-conscious* experience.

[3] Here I am considering a common sense conception of consciousness. I consider different definitions of consciousness and explain what I mean by consciousness in chapter 4.

research, career, politics, ethics, economy, art and so forth – matters to us (in a specific way) by virtue of the fact that we are conscious. To put it the other way around, if we were not conscious, nothing would matter to us. We would not care about anything at all, at least not in any way close to what we usually mean by "caring". Take consciousness away, and we would have no drive or compulsion to care about, invest in, or study anything. We would not taste or cherish life, but rather blindly walk through it. Thus consciousness is a fundamental aspect of life as we know it. It is what makes our life, our activities and our relationships matter to us and be valuable to us. I think that studying consciousness has a deep – almost primordial – theoretical interest because it comes down to studying the phenomenon in virtue of which things matter to us. The pervasiveness and prominence of consciousness in our life make it deservedly a more than worthy object of study.

1.1.2 An Intellectual Challenge

A second reason to take consciousness seriously is that better understanding consciousness is amongst the few remaining great intellectual challenges facing humanity together with understanding the origin of the universe, and a fistful of others. Any such challenge is inherently stimulating because it tackles fundamental questions about the nature of mankind and the universe – philosophical questions concerning our origin and our nature that everybody wonders about at some point. The lack of straightforward solutions to these challenges should not be seen as something frustrating and off putting, but rather – on the contrary – as a motivation fostering the spreading and development of innovative thinking strategies, as well as new theoretical and practical paradigms. Facing big intellectual challenges such as the nature of consciousness, regardless of the adequateness of the resulting answers, is a culturally and intellectually enriching process. I take this to be a second good reason to persevere and keep on confronting the titan (even if just with wooden spoons).

1.1.3 Practical and Ethical Applications

A third reason to take consciousness seriously is that a better understanding of consciousness – what it is and how it works – besides having great theoretical interests would also have an impact in terms of practical gains and applications in everyday life. For example, the mapping of neural correlates of consciousness (see section 1.3) is an important first step toward the medical identification and treatment of dysfunctions of consciousness. Moreover, if it were possible to clearly identify whether (or when, or in which way) a specific individual is conscious, many problematic and controversial issues related to the divide between conscious and non-conscious beings in medicine and ethics would dissolve or at least be better informed. If consciousness and its neural underpinnings had no more secrets, handling thorny topics such as abortion, euthanasia, animal rights, and so on would be at least a little easier. This could also have consequences in terms of legal resolutions and attributions of moral responsibility. Furthermore, if we could understand what makes one conscious and if it were possible to reverse-engineer the process, we could envisage making conscious what is naturally not so. For example, we could engineer conscious machines – machines that would not only pass the Turing test for intelligence, but also feel.

I am not claiming that an understanding of consciousness sufficient to reach such futuristic goals is in sight as yet. As a matter of fact we might well never get that far (that might be either a disappointment or a relief to science fiction fans). Nonetheless, at least in principle, and at least to some – more modest – extent, a better understanding of consciousness can have concrete repercussions (such as informing new developments in medical procedures and legal regulations) that could add on to the already remarkable theoretical benefits of better understanding consciousness. This, I think is a further reason to take consciousness seriously.

1.1.4 A Hopeful Research Project in Philosophy and Science

A fourth reason to take consciousness seriously is that nowadays consciousness is not of interest only to armchair philosophers, but also to scientists. Current scientific consciousness research is successful and its future potential is promising, even if – for contingent reasons – that has not always been the case. For a long time scientists avoided tackling consciousness. As Dehaene puts it:

> Throughout the nineteenth and twentieth centuries, the question of consciousness lay outside the boundaries of normal science. It was a fuzzy, ill-defined domain whose subjectivity put it forever beyond the reach of objective experimentation. For many years, no serious researcher would touch the problem: speculating about consciousness was a tolerated hobby for the aging scientist. (Dehaene 2014, p. 7)

The philosopher John Searle confirmed this trend by recalling asking a famous neurobiologist about consciousness as a young philosopher and being struck by hearing him reply exasperated "look, in my discipline it's okay to be interested in consciousness, but get tenure first."[4] In short, there was a widely spread conviction that consciousness was not a serious and respectable field of study for a scientist, the reason for this being that it was not clear *how* an ontologically subjective phenomenon such as consciousness could be accurately objectively empirically studied. The scientific rebuff of consciousness is nicely illustrated by fact that, as Dehaene reports, until the late 1980s the word "consciousness" was taboo during lab-meetings in cognitive science. No one was allowed to mention the "C-word":

> The general feeling was that using the term *consciousness* added nothing of value to psychological science. In the emerging positive science of cognition, mental operations were to be solely described in terms of the processing of information and its molecular and neuronal implementation. Consciousness was ill defined, unnecessary, and passé. (Dehaene 2014, p. 8)

The conviction that consciousness was not worth taking seriously in empirical research however was ill founded, and in the late 1980s this became progressively clear. According to Dehaene (2014), the transformation of consciousness from a taboo topic into today's laboratory phenomenon has occurred because of (i) the articulation of a clear and univocal definition of consciousness – partly due to philosophical work disentangling concepts such as wakefulness, attention, access consciousness and phenomenal consciousness, (ii) the discovery that conscious access to a certain pieces of information can be experimentally manipulated in a reproducible manner, and (iii) a new respect for subjective phenomena considered as raw data. The combination of these ingredients, combined and supported with the development of new tools of neuroimaging, allowed the birth and the exponential development of a respectable experimental science of consciousness, projecting the problem of consciousness to the forefront of neuroscientific research.

An important reason for the reluctance to consider consciousness seriously was that for a long time it had been unclear how one could explain which part of brain activity corresponds to purely unconscious and automatic operations underlying mental states (e.g., visual perception), and which part leads to consciousness (e.g., being visually conscious of an object). This changed when scientists started developing ways to manipulate consciousness:

> In the past twenty years, cognitive scientists have discovered an amazing variety of ways to manipulate consciousness. Even a minuscule change in experimental design can cause

---

[4] Intervention of Searle, 2012 Summer School "Evolution and Function of Consciousness," Cognitive Science Institute, UQàM.

us to see or not to see. We can flash a word so briefly that viewers will fail to notice it. We can create a carefully cluttered visual scene, in which one item remains wholly invisible to a participant because the other items always win out in the inner competition for conscious perception. We can also distract your attention. [...] The perceived image, the one that makes it into awareness, and the losing image, which vanishes into unconscious oblivion, may differ minimally on the input side. But within the brain, this difference must be amplified, because ultimately you can speak about one but not about the other. Figuring out exactly where and when this amplification occurs is the object of the new science of consciousness. The experimental strategy of creating a minimal contrast between conscious and unconscious perception was the key idea that cracked wide open the doors to the supposedly inaccessible sanctuary of consciousness. [...] The daunting problem of consciousness was reduced to the experimental issue of deciphering the brain mechanisms that distinguish two sets of trials – a much more tractable problem. (Dehaene 2014, pp. 10–11)

This minimal contrast strategy allows the individuation of brain mechanisms that distinguish trials in which subjects are conscious of an item and trials in which they are not conscious of such an item. This strategy, importantly, relies on the one hand on a specific conception of consciousness as consciousness of a given piece of information[5], and on the other hand on taking subjective introspective reports seriously, as raw data. For example, if during an experiment the experimenter shows an image in the subject's visual field and the subject denies (consciously) seeing it, then the image has to be scored as invisible *for* the subject because, unless the subject is lying, his report indicates the fact that needs to be explained (i.e., the subject's lack of consciousness of that image). Taking subjective introspective reports seriously required totally rejecting the behaviouristic paradigm according to which introspection was of no interest for a purely objective experimental science of psychology. As Dehaene (2014, p. 12) rightly indicates, behaviourism wrongly conflated two distinct issues: introspection as a research method, and introspection as raw data. Of course we cannot trust introspective intuition as a research method indicating us how the world really is. For example, if a subject reports an out of body experience, saying that she "literally" felt that she moved out of her body, we are not warranted to take that as proof that – really – her self, mind or soul "physically" moved out of her body. Nevertheless, in order to understand how consciousness works, we have to trust introspective reports as indicating the mental phenomena that beg for an explanation, i.e., as indicating the conscious experiences that one has[6]. In the above case, for example, we have to take seriously, as raw data, the subject's report that

---

[5] I later argue that this conception of consciousness is too restrictive, but I grant that on the assumption of such a conception Dehaene's account is sound.
[6] I elaborate this point in depth in section 1.2.

she felt *as if* she moved out of her body, and try to explain the reported conscious experience by showing what could rationally and scientifically explain that. The explanation of such raw data could – at least in principle – be that something really did move out of the body, but could also be that the content of out of body experiences can be equally empirically explained as being just an illusion or a hallucination.

Without digging further into the historical details of the scientific revolution in the study of consciousness, we can say that the sciences have become more and more successful in giving us insights into the workings of the conscious mind. Even though there are still many theoretical and practical problems yet to be solved and limitations to what can be studied, there has been undeniable remarkable progress in the technology, approaches used, and results obtained in scientific consciousness research over the past thirty years, and further progress is being made. The future research on neural correlates of consciousness looks particularly promising. The current ever growing presence of science alongside philosophy in the study of consciousness is undoubtedly a positive sign. This partnership indicates that consciousness, besides having preserved its original philosophical fascination, is now also widely accepted as a totally respectable and serious object of scientific research. This leaves little room for the skeptics claiming that consciousness is a mystery that cannot be solved[7]. The tremendous progress made in the philosophical and empirical understanding of consciousness in just thirty years shows that taking consciousness seriously can lead to progress.

1.2 Consciousness and the Scientific Paradigm

Clarifying why it is worth taking consciousness seriously is important, but even more important is to understand what are the philosophical and empirical challenges arising once we *do* take it seriously. An important preliminary challenge in this respect consists in rebutting the wrongheaded belief that there is no way in which "objective" science can study consciousness. I tackle this challenge by distinguishing four notions (epistemic objectivity, epistemic subjectivity, ontological objectivity, and ontological subjectivity) and showing that ontologically subjec-

---

[7] See for example McGinn (1989), who famously claimed that consciousness is a mystery that human intelligence will never unravel. The strong claim according to which consciousness is a "mystery" can be opposed to the weaker claim that consciousness is a "puzzle", that is, a problem that is not insolvable in principle.

tive phenomena can fit in the epistemically objective framework of science (1.2.1). On the basis of this preliminary work I conclude that consciousness intended as an ontologically subjective phenomenon could – at least in principle – be the subject of epistemically objective empirical investigation. Granting this I suggest, however, that the road leading to the achievement of a fully satisfactory scientific understanding of consciousness faces further and possibly more troublesome challenges (1.2.2).

1.2.1 Epistemic Objectivity and Ontological Subjectivity

Natural science aims at *epistemic* objectivity – that is, it aims at objective knowledge of facts about the world[8]. As John Searle puts it, natural science "aims at getting a set of truths that are free of our special preferences and prejudices" (Searle 1997, p. 113 ff). Searle calls "epistemically objective" those statements or judgments that are true or false independently of any prejudices or attitudes on the part of observers. For example, "Hans weighs 80 Kilos" is an epistemically objective statement since its truth or falsity is independent of anyone's personal prejudices or preference – it depends solely on objective facts about the thing in question (i.e., whether Hans does or does not *de facto* weigh 80 Kilos). In other words, the truthmakers determining the truth or falsity of epistemically objective statements are solely the intrinsic features of the world.

Epistemically objective statements or judgments such as "Hans weighs 80 Kilos" or "Rembrandt died in 1669" can to be distinguished from *epistemically subjective* statements or judgments such as "Hans is handsome" or "Rembrandt is a better painter than Ivan Aivazovsky".

> We often speak of judgments as being "subjective" when we mean that their truth or falsity cannot be settled "objectively" because the truth or falsity is not a simple matter of fact, but depends on certain attitudes, feelings, and points of view of the makers and the hearers of the judgment. (Searle 1992, p. 94)

"Hans is handsome" is an epistemically subjective statement because its truth of falsity depends at least partially on the attitudes and preferences of the observer – the statement being true if Hans really appears handsome to the observer, and

---

[8] In subsection 1.5.2 I offer a brief excursus questioning the very possibility of achieving anything like epistemic objectivity. I think that – at least indirectly – knowledge as usually defined depends on epistemic subjectivity, even in natural science. Regardless of this, I agree that epistemic objectivity is a laudable objective.

being false if Hans does not appear handsome to the observer. In short, the truth of epistemically subjective statements is subject-relative. There is no objective truth as to whether "Hans is handsome" or "Rembrandt is a better painter than Ivan Aivazovsky" because the "handsomeness" of Hans or the "betterness" of Rembrandt with respect to Ivan Aivazovsky are not intrinsic features of the world, but rather attributes that are observer relative or observer dependent, that is, attributes that exist only relative to some observer or user.

If we endorse the global supervenience claim – i.e., the claim that worlds that are alike in all physical respects are alike in all mental respects as well (cf. section 2.2.3) – then we could be tempted to claim that statements such as "Hans is handsome" have objective truth-makers in the world, since they are ultimately made true by physical facts (about the environment, Hans, and people thinking about Hans)[9]. I concede that, given global supervenience, a complete knowledge of (ontologically objective) physical facts would be sufficient to claim that specific people judge Hans as being handsome, e.g., that for Anna and Lucy the statement "Hans is handsome" is true. However this only shows that there could be ontologically objective truth-makers for the statement "Hans appears to be handsome to several people" which – notice – is an epistemically *objective* statement[10]. This is an epistemically objective statement because the truth or falsity of the statement is independent of people's attitudes and preferences towards it: it is true if there are people (e.g., Anna and Lucy) to which Hans appears handsome, and false if there are no such people. Importantly, the fact that there could be ontologically objective truth-makers for the statement "Hans appears to be handsome to several people" is not sufficient for the stronger conclusion that epistemically subjective statements such as "Hans is handsome" have themselves *objective* truth-makers in the world verifying them. In other words, being able to objectively verify the statement "Hans *subjectively appears* to be handsome to several people" does not entail being able to verify the statement "Hans *is* handsome". The reason for this is that – contrary to epistemically objective statements – the truth or falsity of epistemically subjective statements such as "Hans is handsome," are not independent of anyone's personal prejudices or preferences (and their supervenience basis), and do not depend solely

---

[9] Michael Esfeld raised this point in personal communication (10 January 2015).
[10] What might create a confusion here, and I will come back to this, is that "Hans appears to be handsome to several people" is an epistemically objective statement about ontologically subjective facts. Thus, the only novelty introduced by the supervenience-based argument would be the idea that we can objectively verify epistemically objective statements about ontologically subjective facts – something I do accept.

on objective facts about the thing in question (i.e., Hans). In the absence of people's beliefs that Hans is handsome (and, accordingly, in the absence of all the objective facts about such beliefs), other things being equal, there would be no truthmaker left verifying the statement "Hans is handsome". Opposite to this, in the absence of people's beliefs that Hans weighs 80 Kilos (and, accordingly, in the absence of all the objective facts about such beliefs), other things being equal, there would nonetheless be a truthmaker in the world verifying the statement "Hans weighs 80 Kilos". In fact, there still would be a fact of the matter that Hans weighs that much (i.e., if you put Hans on a scale – regardless of whether there are people looking at it – the scale would rightly indicate "80 Kg"). No similar mind-independent objective scale can reliably indicate whether or not Hans is handsome.

Since epistemically subjective statements or judgments such as "Hans is handsome" are only *subjectively* true of false, and since natural science aims at getting a set of *objective* truths about the natural world rather than at getting a set of subjective truths about how the world appears to be, science relies exclusively on epistemically objective statements or judgments of the kind "Hans weighs 80 Kilos" (or "Hans appears to be handsome to several people").

Now consider the statement "Hans has a back pain". This is an *epistemically objective* statement, since it is made true by the existence of an actual fact (i.e., Hans having a back pain), and is made false by the non-existence of an actual fact (i.e., Hans not having a back pain). This statement is not *epistemically subjective* because it is not dependent on any subjective stance, attitude, or opinion – it depends solely on a plain fact about reality (i.e., whether or not Hans has a given phenomenal experience). Interestingly, however, the truthmaker of statements such as "Hans has a back pain" has a subjective mode of existence: it is *ontologically subjective* (cf. Searle 1992, p. 94). The statement is true if Hans subjectively feels a back pain, and the statement is false if Hans does not subjectively feel a back pain. There is a fact of the matter as to whether Hans has a back pain, but only Hans can subjectively know what is actually the case because he is the only one that has direct access to his subjective facts (i.e., his own experience). I say that the truthmaker – (the presence of) Hans' pain – is ontologically subjective because it is not equally accessible to any observer: only Hans feels his own pain.

Compare this again to the statement "Hans weighs 80 Kilos". Here the truthmaker of the statement has an *ontologically objective* mode of existence. Anybody can know, at least in principle, whether Hans does or does not weigh 80 Kilos (whether the statement is true or false). You do not have to be Hans in order to know whether it is true that Hans weighs 80 Kilos; you just need a tool to measure weight. Thus, to sum up, weight is an example of an *ontologically objective* property because it is equally accessible to any observer, whereas pain is an example of an *ontologically subjective* property because it is not equally accessible to any observer.

The important point here is that there can be *epistemically objective statements* about both ontologically *objective* facts and ontologically *subjective* facts. "Hans weighs 80 Kilos" is an *epistemically objective* statement about an *ontologically objective* fact (i.e., Hans' weight). "Hans has a back pain", on the other hand, is an *epistemically objective* statement about an *ontologically subjective* fact (i.e., Hans' pain). This shows, as Searle correctly points out, that the *epistemic objectivity* of the method of natural science does not entail the *ontological objectivity* of its subject matter. More precisely, the *epistemic objectivity* of the scientific method does not preclude *a priori* the existence of *ontologically subjective* phenomena (such as painful experiences), and – importantly – does not preclude the possibility of studying *ontologically subjective* phenomena.

Science can, and as a matter of fact usually does, apply its *epistemically objective* method to study *ontologically objective* properties of rocks, molecules, force, mass or gravitational attraction, that is, properties that have an ontologically objective mode of existence – an existence that is equally accessible by any observer (i.e., observer independent). However the fact that natural science works in an *epistemically objective* paradigm should not be taken to suggest that reality is limited to the *ontologically objective* domain, containing no irreducibly *ontologically subjective* phenomena (cf. Searle 1997, p. 114; Searle 1992, p. 19). Furthermore, the fact that natural science works in an *epistemically objective* paradigm should not be taken to suggest that science is limited to the study of the *ontologically objective* domain:

> It is just an objective fact – in the epistemic sense – that I and people like me have pains. But the mode of existence of these pains is subjective – in the ontological sense. [...] One part of the world consists of ontologically subjective phenomena. If we have a definition of science that forbids us from investigating that part of the world, it is the definition that has to be changed and not the world. (Searle 1997, p. 114)

Indeed, science can apply its epistemically objective method to study phenomena like consciousness, pains and emotions – that is, ontologically subjective phenomena that exist despite not being equally accessible to any observer.

So far I have claimed that phenomena that have a subjective mode of existence (i.e., a subject-relative existence that depends on being felt by a subject) can be the subject matter of science. Think for example of headaches. Even though headaches have an ontologically subjective mode of existence (i.e., they exist only insofar as they are felt by someone), there are epistemically objective facts about headaches. For example, it is an epistemically objective fact that people have headaches and that I have a headache now. Furthermore, we can utter epistemically objective statements such as "people have headaches" or "Pietro has a headache now", and the truth value of these statements does not depend on anyone's opinions or attitudes – not even my own; it only depends on plain facts about (subjective) reality. The statement "Pietro has a headache now" is true if and only if Pietro does have a headache. The only thing distinguishing *ontologically subjective* facts from *ontologically objective* facts is their mode of presentation (subjective and objective respectively). Thus, to conclude, there is no reason in principle why the *epistemically objective* method of science should not be applicable to both *ontologically objective* things like rocks and molecules, and to *ontologically subjective* phenomena such as consciousness (e.g., pains, emotions, and so on).

Table 1 summarizes the possible combinations of epistemic statements and ontological facts. In grey are highlighted examples of facts that – at least in principle – can be the object of epistemically objective scientific claims.

Table 1

	Epistemically objective statements	Epistemically subjective statements
Ontologically objective facts	Hans weighs 80 Kilos. Humans have brains.	Hans is handsome. Brains are disgusting.
Ontologically subjective facts	Hans has a back pain. People have desires.	Hans' pain is worse than mine. Desires are wrong.

1.2.2 Dealing With the Subjective Ontology of Consciousness

It is common to have phenomenally conscious experiences such as feeling tedious headaches, feeling cold, hearing loud noises, feeling excited, tasting pizza, feeling angry, feeling tired and countless more. These experiences can be defined *phenomenally* conscious because there is *something it is like* to have them, phenomenologically, something it is like *for* the being (cf. Nagel 1974). Conscious experiences so intended have a *subjective ontology* – they are ontologically subjective phenomena – because they have an ontologically subjective mode of existence: they exist only insofar as a subject has conscious experiences and they exist only from the subjective (or first-person) point of view of that subject – i.e., they exist only *for* the subject, they are always *someone's* conscious experience. If we want to give an all-embracing account of reality it is fundamental to acknowledge the fact that consciousness – any phenomenally conscious experience or mental state – has a subjective (or first-person) ontology. A passage by John Searle emphasizes effectively the importance of coming to terms with the ontological subjectivity of consciousness:

> It would be difficult to exaggerate the disastrous effects that the failure to come to terms with the subjectivity of consciousness has had on the philosophical and psychological work of the past half century. In ways that are not at all obvious on the surface, much of the bankruptcy of most work in the philosophy of mind and a great deal of the sterility of academic psychology over the past fifty years, over the whole of my intellectual lifetime, have come from a persistent failure to recognize and come to terms with the fact that the ontology of the mental is an irreducibly first-person ontology. (Searle 1992, p. 95).

The reference to the "irreducibility" of the first-person ontology could be interpreted in two distinct manners. On the one hand, it could be interpreted as suggesting that the ontologically subjective properties of conscious experience are not *epistemically reducible* to ontologically objective properties of matter, i.e., knowing everything about the ontologically objective properties of matter does not suffice to know everything about ontologically subjective properties of conscious experience. On the other hand, it could be interpreted in a stronger sense as suggesting that ontologically subjective properties of conscious experience are not *ontologically reducible* to ontologically objective properties of matter, i.e., that ontologically objective properties of matter do not exhaust everything that there is. I interpret Searle as defending the latter – stronger – reading. The issue that Searle raises is that the ontologically subjective mode of existence of consciousness is distinct and not reducible to the ontologically objective mode of existence of bodies, brains and the like (I return to these issues at length in section 1.4, and in chapter 2).

How does this observation affect the way in which consciousness can be studied? As we have seen in section 1.2.1, the fact that consciousness is *ontologically subjective* is not in principle incompatible with the *epistemically objective* approach of science. A scientific account of consciousness is possible. However, it should be clear that the science of consciousness is bound to explaining *epistemically objective* facts about *ontologically subjective* phenomena. Whether explaining this is sufficient to achieve a *fully satisfactory* understanding of consciousness is an open question.

Even if there is no reason in principle why "objective" science could not study consciousness intended as *ontologically subjective* phenomenon, it would be silly to deny that, in practice, the path leading to a fully satisfactory scientific understanding of consciousness is a tough one, filled with troublesome challenges and – according to some – hiding a dead end. On the one hand there are explanatory worries concerning what the science of consciousness can tell us about consciousness, and methodological questions concerning how we could empirically study consciousness due to the mode of existence of the *explanandum* (see section 1.3). On the other hand there are philosophical worries regarding the limits inherent in any scientific explanation of consciousness (see section 1.4). These, I claim, are the relevant challenges arising when we try to fit consciousness in the scientific paradigm.

The explanatory and methodological worries are due to the fact that consciousness, intended as *ontologically subjective* phenomenon, is not equally accessible to any observer. Unlike any ontologically objective natural phenomena that science usually deals with, because of its subjective mode of existence, consciousness is exceptionally resilient to epistemically objective descriptions, measuring, framing or quantification. This poses a series of problems for the scientist aiming at giving an epistemically objective account of consciousness. First, it is not clear how *explanatory* a scientific account of consciousness can be. Second, it is not methodologically clear *how* one should empirically proceed to gather epistemically objective data about a phenomenon whose existence is intrinsically tied to an individual, first-person, ontologically subjective perspective. The path to a satisfactory scientific understanding of consciousness requires developing strategies to overcome or circumvent these limitations.

The science of consciousness aiming at individuating the Neural Correlates of Consciousness is an example of a scientific approach to consciousness that has effectively developed despite having to embrace these unusual setup conditions (see section 1.3). The results obtained by relying on phenomenological subjective reports as raw data indicating what has to be empirically explained and developing ways to study neural changes correlating with changes in consciousness are astonishing. Nevertheless, it has to be noted that NCC studies are explanatorily limited. Furthermore, it is not always clear whether and to which extent experimental designs in NCC research manipulate (even if unwillingly) subjective reports, nor whether the conclusions of the studies really are conclusions about phenomenal consciousness rather than about something else (e.g., access consciousness). In sum, even if there certainly are praiseworthy attempts to empirically study consciousness despite its subjective dimension, there are still explanatory and methodological worries that cause trouble in evaluating whether and to which extent these attempts are fruitful and accurate.

In addition to these worries, another relevant challenge arising when trying to fit consciousness in the scientific paradigm is the one posed by philosophers claiming that any scientific explanation of consciousness has inherent limits to *what* it can explain (see section 1.4). According to some philosophers, even if we might be able to scientifically understand some things about consciousness, there will always be an unbridgeable explanatory gap preventing us to fully explain everything there is to know about consciousness in the epistemically objective language of science. In particular, the claim is that there is no way to scientifically explain why consciousness is as it is – i.e., why it feels like it feels –, or why it exists at all, solely on the basis of epistemically objective facts about it. If these arguments are sound, all we can objectively know about consciousness is not jointly sufficient to explain all there is to know about consciousness, and what remains inevitably beyond the scope of empirical research is precisely what makes consciousness so special: its ontologically subjective dimension.

Summarizing, I suggested that consciousness is a fundamental component of reality characterized by the peculiarity of having an *ontologically subjective* mode of existence. Even though there is no reason in principle why science could not study consciousness intended as an *ontologically subjective* phenomenon, a full understanding and description of consciousness in terms of mind-independent, epistemically objective facts still faces two kinds of worries. On the one hand there are explanatory and methodological problems linked to the *ontologically subjective* mode of existence of the *explanandum*, making consciousness slippery, hardly traceable, and problematic to track down satisfactorily to the vocabulary and general practices of the natural sciences. On the other hand there are philosophical worries regarding the inherent limits of any scientific explanation of consciousness. I explore these problems more in depth in sections 1.3 and 1.4 respectively.

1.3 Neural Correlates of Consciousness (NCCs)

So far I have argued that it is worth taking consciousness seriously (section 1.1.) and I introduced some important challenges that arise once we *do* take it seriously as an *ontologically subjective* phenomenon subject to *epistemically objective* investigation (section 1.2.). In the present section I explain what the quest for the Neural Correlates of Consciousness consists of (section 1.3.1), and I discuss in some detail what the limits of such a research approach to consciousness are (section 1.3.2).

I take it to be uncontroversial and perfectly reasonable to hold that our conscious experience is somehow connected with what goes on in our body and – more specifically – in our brain. That is, I think it is reasonable to hold that conscious experience is not entirely independent and unrelated to what happens in our body or brain. This conclusion is derived from the empirical observation that any ontologically subjective change in one's consciousness (i.e., changes in *whether* one is conscious, or in *what* one is conscious of) consistently correlates with some ontologically objective change in brain activity[11]. Moreover, alterations at the neural level can be accompanied by changes in terms of consciousness. For example, brain damage can be accompanied by drastic changes in the person's consciousness, stimulation of specific neural areas can be accompanied by the appearance of specific conscious experiences, and the assumption of drugs that affect brain functions can be accompanied by changes in conscious experience. Given this, on the assumption that some sort of correlation between consciousness and brain activity exists, it is natural to wonder what these correlations are and how they work.

The most successful and prominent contemporary approach in the science of consciousness is geared precisely toward answering these questions by identifying the Neural Correlates of Consciousness (henceforth NCCs). The expression "neural correlate of consciousness" was probably first used in print by Francis Crick and Christof Koch (1990), and has quickly become a very popular expression used across many disciplines involved in empirical consciousness research. The standard epistemic goal of correlation studies regarding consciousness consists in isolating the minimally sufficient neural correlate for specific kinds of phenomenal content, that is, the minimal set of properties, described on an appropriate level of neuroscientific analysis, that is sufficient to activate a given conscious mental content in a given organism (cf. Metzinger 2000, pp. 3–4). As the neuroscientist Christof Koch says:

> We do assume that any phenomenological state – seeing a dog, having pain, and so on – depends on a brain state. The neuronal correlates of consciousness are the minimal set of neuronal events jointly sufficient for a specific conscious phenomenal state (given the appropriate background enabling conditions). Every [conscious] percept is accompanied by some NCC. (Koch 2004, p. 304)

My goal here is not to discuss specific studies, results or proposals for actual NCCs[12]. That is a task for neuroscientists. I am interested in understanding what

---

11 I elaborate this in much more detail in section 2.2.3 by introducing the notion of supervenience. Notice, importantly, that all I am claiming is that there is a *correlation* between conscious experience and neural activity. I am not committing to any causal claim (i.e., what causes what).
12 Chalmers (2000, p. 17) lists a few candidate NCCs: 40-herz oscillations in the cerebral cortex (Crick and Koch 1990), extended reticular-thalamic activation system (Newman and Baars 1993), 40-herz rhythmic activity in thalamocortical systems (Llinás et al. 1994), intralaminar nuclei in the thalamus (Bogen 1995), neural assemblies bound by NMDA (Flohr 1995), neurons in the extrastriate visual cortex projecting to prefrontal areas (Crick and Koch 1995), visual processing within the ventral stream (Milner and Goodale 1995), specific neurochemical levels of activation (Hobson 1997), and specific neurons in the inferior temporal cortex (Sheinberg and Logothetis

NCC research is geared toward, and discussing what are its limits and what it can tell us about consciousness.

1.3.1 Defining NCC

Without going into technical details, it is interesting no notice that there are different ways of defining NCCs[13]. Here is how David Chalmers (2000, p. 18) tentatively defines NCC:

> *def.1:* A neural system N is an NCC if the state of N correlates directly with states of consciousness.

There are several ways to interpret what is meant by "states of consciousness". First, one could look for the NCC of *creature consciousness*, roughly, the property a creature has when it is conscious, and lacks when it is not conscious[14]. The dominant view under this interpretation is that the NCC of creature consciousness is located in or around the thalamus, or at least that it involves interactions between the thalamic and cortical system. Instead of looking for the NCC of creature consciousness, one could look for the NCC of the background state of consciousness, an overall state of consciousness such as being aware/asleep, dreaming, being under hypnosis, and so on. The location of such states of consciousness seems to match the one just described (Chalmers 2000, p. 18). As we already said, however, NCC research is mostly concerned with the NCC of the *contents* of consciousness, even if this is often tacit in empirical studies (e.g., Milner and Goodale 1995). In particular, most of the work on *visual* NCCs – and, according to Chalmers (2000, p. 19), much of the most interesting work – is concerned precisely with the NCC of the content of specific visually conscious states. This sort of research presupposes that conscious mental states (e.g., conscious visual perceptions) have some sort of specific content representing the world as being one way or another, and investigate what kind of minimal neural process correlates with the conscious content at stake. Chalmers defines the latter kind of NCCs as follows:

1997). See (Crick and Koch 1998; Milner 1995) for reviews on NCCs – in particular NCC of visual consciousness. See Chalmers (1998) for a longer list.
13 This subsection is substantially based on the work by David Chalmers (2000, pp. 18–39).
14 I introduce the different notions of consciousness in subsection 4.1.2. As it will become clear, I am primarily interested in this sense of consciousness.

def.2: A neural correlate of the contents of consciousness is a neural representational system N such that representation of a content in N directly correlates with representation of that content in consciousness.

or, in short:

def.3: A content NCC is a neural representational system N such that the content of N directly correlates with the content of consciousness.

For any distinctive kind of conscious experience, there is a corresponding phenomenal property, namely the property of having a conscious experience of that kind (with a specific *quale*). For example, having a conscious visual perception of vertical lines is a kind of phenomenal property that differs from the phenomenal property of feeling sharp pain. Chalmers (2000, p. 22) defines the neural correlate of an arbitrary phenomenal property P accordingly:

def.4: A state N1 of system N is a neural correlate of phenomenal property P if N's being in N1 directly correlates with the subject having P.

That is, for example, a specific state N1 of the neural system is an NCC of feelings of sharp pain if the fact that the neural system is in state N1 directly correlates with a subject feeling sharp pain.

Usually, however, NCC studies are concerned with neural correlates of families of phenomenal properties rather than with neural correlates of specific phenomenal properties, where a phenomenal family is defined as a set of mutually exclusive phenomenal properties that jointly partition the space of conscious experiences, or at least some subset of that space. For example, work on the visual NCC is concerned with the NCC of the whole system of visual contents – not just vertical lines.

def.5: A neural correlate of a phenomenal family S is a neural system N such that the state of N directly correlates with the subject's phenomenal property in S.

In other terms, an NCC of the phenomenal family "conscious visual perception" is a neural system N such that the state of the neural system N directly correlates with the subject's phenomenal property in (any) conscious visual perception.

If we apply this definition to the NCC of creature consciousness we would need a neural system with two states correlating with the two properties of a simple phenomenal family – being conscious and not being conscious. If we apply this definition to the NCC of background states of consciousness we would need a neural system with a few states correlating directly with the few properties of a phenomenal family (e.g., dreaming, waking state, hypnosis, and so on). However, if we apply this definition to the NCC of the contents of consciousness, we would need a neural system with many states correlating directly with the many properties of the phenomenal families (e.g., particular beliefs, desires, visual perceptions, and so on.). This requires that the content of the neural state match the content of consciousness systematically.

In all the definitions of NCC considered so far we said that the neural state correlates "directly" with states of consciousness. But how strong a relation is required for a neural system N to be an NCC? Chalmers argues (2000, p. 25) that an NCC has to be a *minimal sufficient system*, that is, a minimal system whose state is sufficient for the corresponding conscious state. By this definition, a neural state N will be an NCC when (i) the states of N suffice for the corresponding states of consciousness, and (ii) no proper part M of N is such that the states of M suffice for the corresponding states of consciousness. Note that on this definition there may be more than one NCC for a given conscious state type.

Since to say that a neural state N1 suffices for conscious state C is to say that this occurs with respect to a range of cases (i.e., across very many individuals), it is reasonable to ask over what range of cases the correlations are individuated and must hold. Chalmers considers a number of possible situations in which NCCs could be studied (i.e., ordinary functioning brain in ordinary environments, normal brain and unusual inputs, normal brain and varying brain stimulation, abnormal functioning due to lesions) and concludes that one has to be very cautious when drawing conclusions about NCC location from lesion studies. He claims that the lesion in question might change the brain architecture in such a way that what was once an NCC prior to the lesion is no longer an NCC (Chalmers 2000, p. 31). If it is true that depending on the situation one can have a given neural state with or without the corresponding conscious state, then the correlation "system in state N1 – subject having conscious state C" might not range over all possible situations, and the NCC should be individuated accordingly. Chalmers offers a definition of NCC that takes into account the conditions for correlation. Conditions C might be seen as conditions involving normal brain functioning, allowing unusual inputs and limited brain stimulation, but not lesions or other changes in architecture:

> *def.6:* An NCC is a minimal neural system N such that there is a mapping from states of N to states of consciousness, where a given state of N is sufficient, under conditions C, for the corresponding state of consciousness.

Accordingly, for the central case of the neural correlate of the content of consciousness:

def.7: An NCC (for content) is a minimal neural representational system N such that representation of a content in N is sufficient, under conditions C, for representation of that content in consciousness.

These two definitions of NCC are coherent, well motivated in their own right, and fit the way the notion is generally used in the field, making sense of empirical research. I suggest considering NCCs so defined.

1.3.2 Explanatory Limits and Fuzziness of NCC Research

The empirical research project geared toward the search for NCC of content is effective, successful, and flourishing in neuroscience laboratories all over the world. A reason for the popularity and enthusiasm surrounding this kind of research is that now that the technical means and the core methodology for an effective epistemically objective study of consciousness are available, they can readily be applied – with only minor adjustments – to explore and study a vast variety of conscious phenomena that were previously unattainable. NCC research has produced a lot of interesting data and hugely contributed to the understanding of mental phenomena in the past decades. Nevertheless, besides mentioning the positive effects of such research, it is worth also questioning the limits of such an approach. In particular, I believe it is important to consider *what* this kind of research is ultimately telling us about consciousness, and to question some of the underlying methodological presuppositions leading to the empirical correlation results. In fact, even though the rough principle guiding correlation research is clear and its goal is laudable, the interpretation and individuation of correlations between an ontologically subjective phenomenon (phenomenal consciousness) and an ontologically objective state of a physical system (a brain state) are delicate matters deserving careful philosophical attention.

Chalmers (2000, p. 37) argues that an NCC (i) is not necessarily a system solely or mainly dedicated to consciousness, (ii) it might not be the brain system most responsible for the generation of consciousness, (iii) it might not yield an explanation of consciousness, and (iv) identifying an NCC might not be the key to understanding the processes underlying consciousness. If these claims are correct, then it appears that NCC research might have strikingly disappointing explanatory limits.

The most important limit of NCC research in my opinion does not come as a surprise, given the explicit and well-advertised limitation of NCC research to the study of *correlations*. The limit of correlations studies is that, even if successful, they do not suffice to explain *why* or *in virtue of what* the correlations individuated hold.

Any consistent, accurate and reliable correlation between two events or phenomena X and Y can have four sorts of explanations:
1) X causes Y
2) Y causes X
3) X and Y are both caused by something else (Z)
4) X and Y are actually the same thing (they are identical)

For example, the observable correlation of the heart pumping frequency (X) with the frequency of the heartbeat sound (Y) is explained by the fact that the blood pumping of the heart causes the heartbeat sound (X causes Y, and all the other options are erroneous). On the other hand, the correlation of changes in the entropy of water (X) with changes in the molecular structure of H_2O (Y) is explained by the fact that water is identical with H_2O (X is identical to Y, and all the other options are erroneous). Interestingly, however, the mere observation of these correlations is not by itself sufficient to conclude how the correlation is to be explained.

Going back to the case of NCC research, even though this research might highlight correlation between neural states (N) and conscious states (C), we are left with four possible ways underlying and explaining such correlation:
1) N cause C
2) C cause N
3) N and C are both caused by something else (...but what?)
4) N and C are actually the same thing (they are identical)

The quest for NCCs only identifies the minimal ontologically objective neural system that correlates with a specific ontologically subjective conscious experience. NCC research alone, therefore, does not tell us which of the above options accurately explains the correlations observed. More precisely, finding NCCs does not come down to explaining what causes what (i.e., which among options 1, 2 and 3 is right), nor that consciousness can be *reduced* or *identified* (ontologically) to brain activity (option 4). Thus, importantly, the quest for NCCs falls short of empirically elucidating what drives and explains the correlation. The interpretation of the correlations is a separate issue, and this is where theorizing – both scientific and philosophical – comes into play. This being said, the explanatory modesty (or lack of explanatory ambition) of NCC studies is not entirely a bad thing. Indeed, it surely is better to uncover epistemically objective correlations between consciousness and neural activity than to know nothing at all.

In addition to the explanatory limits of NCC studies, however, I also have doubts regarding some of the standard methodological assumptions driving this research and leading to the supposedly epistemically objective conclusions. In particular, I find disconcerting that the results obtained in empirical testing constrained by strong contingent limitations are sometimes taken at face value and used to imply conclusions extending well beyond what the experiments actually prove.

NCC studies usually proceed by means of a "minimal contrast" method (or contrastive method) consisting in (i) creating a setup in which a subject can be either conscious or not conscious *of* a given target, and (ii) investigating what neural changes accompany the reported changes in consciousness *of* the target under consideration. A first thing to notice here is that the notion of consciousness considered is a transitive one (i.e., consciousness *of* a specific target). Focussing only on one's consciousness (or lack of consciousness) *of* a specific stimulus has the practical advantage of allowing the scientist to focus on correlating exclusively that change. For example, a researcher interested in tracing the neural difference between case (i) in which one is visually conscious *of* a specific target stimulus in her visual field, and case (ii) in which one is not visually conscious *of* that target stimulus in her visual field, will have to trace only two distinct sets of trials. The gathered data will supposedly indicate what neural difference distinguishes the set of trials where the subject is conscious of the target from the set of trials where the subject is not conscious of the target. However, are we sure that the correlation really is only a correlation of *that* change in consciousness? It should not be forgotten that the experiment consists in a lab set-up forcefully having to abstract from a whole lot of contextual information – and that this impacts the results. There might be (and usually is) more to the subject's consciousness than – say – only the presence or absence of visual consciousness *of* the target stimulus. I find doubtful the assumption that the neural activity considered in the study correlates solely with the visual consciousness *of* that specific stimulus – it correlates with whatever (everything) the subject was conscious of (as well as many other things). Of course, the more such an experiment is reproduced the less coarse-grained the result will be, but this does not yet mean that the result will be crystal clear.

My complaint is that whenever empirical studies firmly partition one's overall conscious experience in watertight compartments – as when they focus only on one's visual consciousness of a given target – they seem to be tacitly assuming that one's conscious experience can be "dissected" into independent parts and analysed one piece at a time depending on what is of interest. But this seems to be a totally unnatural, unwarranted, and perhaps even illusory enterprise. Phenomenologically speaking, we are very rarely (if ever) only visually conscious, or only auditorily conscious. Even though we can make sense of the idea that sense modalities are distinct, the raw data that has to be considered is experience as it is phenomenologically – not what we think about how experience is composed.

Whenever scientists focus only on the NCCs of visual consciousness or auditory consciousness they are artificially imposing rigid phenomenological boundaries that, I claim, are not warranted by our everyday natural conscious experience. By assuming that visual consciousness can be isolated and studied independently of other kinds of consciousness for the sake of empirical simplicity, scientists are *de facto* ignoring the natural overall phenomenological architecture of consciousness (i.e., its unity) – and thus are looking for correlations of a phenomenon that does not exist as such in nature. For this reason I fear that the NCCs individuated by focusing on modality specific conscious mental states can only be indicatively right, and that they might not be the precise neural correlates that empirical studies conclude they are. It is not my intention to argue here against the methodological assumptions of empirical consciousness research since this would require a much more careful analysis than I can afford in the present work. Still I find it important to at least highlight the fact that some of the methodological assumptions might be controversial.

To conclude, even though scientific NCC research is useful and extremely interesting, it is confronted with at least two issues. First, there are serious limits to the explanatory power of correlation studies. Second, it is unclear whether NCCs can be methodologically individuated as they usually are, and it is difficult to understand what such correlations are exactly correlations of. Thus, even though it is certainly important to keep on studying consciousness scientifically by means of NCCs, we should not take for granted the fact that the science of consciousness is already sufficiently equipped to provide us with a fully satisfactory account of consciousness (or even just of correlations).

1.4 The Hard Problem of Consciousness

> [H]ow it is that anything so remarkable as a state of consciousness comes about as a result of irritating nervous tissue, is just as unaccountable as the appearance of the Djinn, when Aladdin rubbed his lamp (Huxley 1866)

As we have seen in the previous section, the project of fitting consciousness in the scientific paradigm, as in the example of NCC research, raises explanatory and methodological worries. It might be tempting to think that, even though there presently are such worries, this is just a contingent fact due to the relatively young stage of development of scientific consciousness research. After all, thinking optimistically, we could imagine that further developing and perfecting the methodology and the strategy currently adopted to study consciousness might suffice to avoid many such worries and to render the results obtained by consciousness research richer and uncontroversial. The trouble, however, is that lurking underneath the minor, contingent, and perhaps partially solvable worries that rise to the surface there seems to be a much deeper, troublesome, and perhaps unsolvable philosophical problem for the scientific study of consciousness. This problem consists in questioning *how* science could ever give a fully satisfactory account of consciousness, given that any epistemically objective account of consciousness is inherently bound to fall short of explaining some of its most crucial aspects.

Most philosophers would subscribe to the thesis that phenomenal content, in a strong sense, supervenes on properties of the brain (Metzinger 2000, p. 2). In other words, most would agree that as soon as all the properties of your brain are fixed, all properties of your conscious experience (i.e., what it is like to be in a given conscious state for you) are fully determined as well. As we have seen, NCC research supports this thesis. However, it is hard to understand *why* phenomenal consciousness in general, and specific phenomenal experiences of one kind rather another, emerge as they do in a self-organizing physical universe correlating, for example, with neural activity of a given kind[15]. The problem known as "Hard Problem" of consciousness consists in the fact that no epistemically objective explanation of consciousness given wholly in physical terms can ever account for – explain – the emergence of conscious experience (cf. Chalmers 1996, p. 93). The worry is that however accurate the explanation of the physical processes correlating with (and supposedly underlying) our conscious experiences is – something that NCC research might yield –, there will always be a further question, namely: why are these processes accompanied by conscious experience? As Chalmers puts it:

> Once we have explained all the physical structure in the vicinity of the brain, and we have explained how all the various brain functions are performed, there is a further sort of explanandum: consciousness itself. Why should all this structure and function give rise to experience? (Chalmers 1996, p. 107)

Chalmers is suggesting that consciousness can be studied scientifically, but that the achievements of scientific studies of consciousness are bound to answering the so-called "easy problems" (Chalmers 1995), that is, problems that – however *technically* complex – could be handled by the standard methods of cognitive science. The "easy problems" consist in explaining how given conscious func-

---

[15] Notice that the Hard Problem is different from the problem of explaining how given surface properties emerge from more fundamental physical properties (e.g., how liquidity emerges in agglomerates of H_2O molecules, or how colour emerges from colourless atoms). I come back to this in section 2.1.3.

tions such as the ability to discriminate, categorize, and react to environmental stimuli, the integration of information by a cognitive system, the reportability of mental states, the ability of a system to access its own internal states, the focus of attention, the deliberate control of behaviour, or the difference between wakefulness and sleep are realized. NCC studies are definitely an important tool leading to the solution of these easy problems. Importantly, however, it is not clear how correlation studies (or any other approach in the science of consciousness, for that matter) could help in answering the Hard Problem, the question of knowing "why" neural states correlate with consciousness, or with specific conscious states. The "hardness" of the Hard Problem consists in the fact that we cannot even conceive of a methodology that could be adopted to successfully explain how subjective Technicolor phenomenology emerges from soggy grey matter – to mention a famous expression by Colin McGinn (1989, p. 349). As neuroscientist Christoph Koch writes, "why qualia *feel* the way they do remains an enigma" (2004, p. 310, original emphasis).

The Hard Problem is widely acknowledged as a serious philosophical problem for a scientific understanding of consciousness and has had a tremendous impact on consciousness literature, even if not everyone agrees that there is such a problem[16]. The tension between the defenders and the challengers of the Hard Problem of consciousness is traceable at many levels and is particularly decisive for the development of opposing philosophical positions concerning the mind-body problem (see chapter 2). Even if it is in principle possible to conciliate consciousness and the scientific paradigm, it is fair to say that, unless the Hard Problem of consciousness is satisfactorily solved or dismissed, there are good reasons to regard consciousness as something of a puzzle – an indomitable[17] natural phenomenon resisting the dominant and otherwise prosperous explanatory strategy adopted by science.

Assuming that there is a Hard Problem, how could one solve it? The most straightforward way could be to (i) show that it *is* logically necessary that consciousness is globally supervenient on the physical – that is, that all the micro-

---

[16] During the 2014 Towards a Science of Consciousness conference in Tucson, Arizona, David Chalmers asked the interdisciplinary audience (approximately 300 people) to raise their hand if they believed there is a Hard Problem. Roughly two thirds of the audience did. This suggests that some people, but not everyone, agrees that there is a Hard Problem. In some cases this may be a consequence of confusion, but in other cases this is a genuine and fundamental disagreement on what are the problems of consciousness. Most of such arguments are based on the idea that there is nothing to be reductively explained (see for example Dennett 1991).

[17] The "indomitable" metaphor is inspired by the "irréductibles Gaulois" holding against the Roman invaders in the Asterix comics by R. Goscinny and A. Uderzo.

physical facts in the world *do* entail *a priori* the facts about consciousness/experience (phenomenal facts). Alternatively, one could try to (ii) show that, even though it is *not* logically necessary that consciousness is globally supervenient on the physical, that does not entail – as Chalmers argues – that no explanation given wholly in physical terms can ever account for the emergence of conscious experience. In other terms, one could try to show that (at least) some explanation given wholly in physical terms can account for the emergence of conscious experience regardless of whether consciousness is (or is not) logically globally supervenient on the physical. Finally, one could try to (iii) explain *why* physical processes are accompanied by conscious experience. The challenge does not consist in demonstrating *that* physical processes are accompanied by conscious experience (a matter of empirical observation), but rather in producing an explanatory theory giving reasons explaining *why* this happens, i.e., what makes it so that physical processes are accompanied by conscious experience. If none of these strategies turns out to be successful, the Hard Problem would remain unsolved[18].

A pragmatic person may claim that knowing *why* physical processes are accompanied by conscious experience (i.e., solving the Hard Problem) is not as important as knowing *how* physical processes are accompanied by conscious experience (i.e., solving the easy problems). Pragmatically speaking, the know-how is more urgent since it could have immediate practical applications. Of course it is fine to aim at is this sort of progress, and in this sense I agree that the priority on the research agenda would have to be set on understanding *how* physical processes are accompanied by conscious experience. As long as our scientific theories on how physical processes are accompanied by conscious experience are accurate enough to work in predicting which kind of conscious experience correlates with such and such brain state (e.g., by means of NCCs), knowing *why* this is the case seems to be of secondary importance. This, however, does not amount to showing that there is no Hard Problem of consciousness and does not affect the main point at stake.

It is time for a summary. After having explained why I think it is worth studying and asking philosophical questions about consciousness in the first place (section 1.1), I have argued that there is no reason in principle why science could not study consciousness intended as *ontologically subjective* phenomenon (section 1.2). Indeed, the science of consciousness – for example by means of NCC

---

[18] In the present dissertation I am not directly or mainly concerned with the Hard Problem. Nevertheless this is a major issue in consciousness studies and thus it is good to have it in mind while tackling other problems.

research – has shown great potential in giving satisfactory epistemically objective accounts of consciousness. Nevertheless, because of the subjective mode of existence of consciousness, there are reasons to think that science cannot go *all* the way in explaining consciousness. On the one hand there are explanatory and methodological worries regarding what exactly is currently being studied and how consciousness is being studied – for example in NCC research (section 1.3). On the other hand, importantly, there is a Hard Problem of consciousness raising a deep philosophical worry regarding the possibly inherently limited explanatory scope of any account of consciousness developed within scientific practice (section 1.4).

1.5 Consciousness: Toward A Diachronic Approach

Both the NCC research and the Hard Problem of consciousness tackle consciousness in a synchronic dimension. I want to show that this way of thinking about consciousness covers only part of the problems of consciousness, and that by considering also the diachronic dimension to which consciousness belongs new interesting questions about consciousness arise[19]. I begin by showing in which sense the ways of thinking about consciousness considered so far are synchronic and by raising some examples of diachronic questions about consciousness. Granting that we have an epistemically limited access to the history of consciousness, I propose a thought experiment suggesting a possible historical scenario and on that basis I raise some diachronic questions about consciousness (1.5.1). I conclude the chapter with a brief excursus on the first-person/third-person distinction (1.5.2).

The contemporary science of consciousness works in a synchronic paradigm. The quest for NCCs, for example, consists in looking for specific neural signatures that are (synchronically) concurrent and correlated with specific conscious states[20]. An NCC, we have said, is a minimal neural system N such that there is a

---

[19] Notice that my point here is not related to the question of whether one's history plays a role in determining whether one is or is not conscious at any given time. That is, I am not trying to claim anything about consciousness along the lines of what teleosemantics suggests about intentional content, i.e., that the content of intentional states depends on history (e.g., that the content of swampman's beliefs differs from the content of Davidson's beliefs because they have a different history; Davidson 1987). I am only claiming that consciousness – as a general phenomenon – has a history, and that it is interesting considering it. I thank Michael Esfeld for driving my attention to this possible source of confusion.
[20] The actual temporal relation between the objective appearance of a given neural state and

mapping from states of N to states of consciousness, where a given state of N is sufficient, under conditions C, for the corresponding state of consciousness. This is a synchronic view: if – *at t1* – there is a minimal neural system N such that there is a mapping from states of N to states of consciousness, where a given state of N is sufficient, under conditions C, for the corresponding state of consciousness, *at t1* there are also the corresponding states of consciousness. More plainly, if at *t1* one has the neural correlate of pain, it has the correlated phenomenally conscious experience of pain also at *t1*. The synchronic temporal dimension of correlations is implicit in the term "correlation" and this explains why it usually does not explicitly appear in definitions of NCCs such as the ones we have previously considered.

The Hard Problem of consciousness – the problem of explaining *why* a given system is conscious or *why* it is conscious in the way it is – is a synchronic problem as well. I call it "synchronic" because it questions the synchronic metaphysical relation between some ontologically objective features of reality (i.e., neural states) and some ontologically subjective features of reality (i.e., conscious mental states).

Much has been said about consciousness in the synchronic dimension. Surprisingly, however, the diachronic dimension of consciousness has not received nearly as much attention in the philosophy of mind. More than that: it has been vastly ignored. I find this very interesting and stimulating. Thinking about consciousness from a diachronic perspective may enable us to uncover interesting philosophical questions that in the synchronic paradigm would otherwise remain unasked. Furthermore, considering consciousness in the temporal dimension may offer a different perspective on already existing problems, allowing us to enrich the discussion with further elements.

In addition to asking how and why physical processes *synchronically* correlate with conscious experience – that is, in addition to synchronic questions – we could ask also *diachronic* questions ("when" questions) about consciousness and its neural correlates. To begin with, we could ask *when* conscious experience of a specific kind (C) and the neural states correlating with that conscious experience first appeared in phylogenetic history, and *when* they appear in the ontogenetic history of specific beings. That is, we could ask the following questions:

---

the subjective appearance of the correlated conscious phenomenal experience may be *quasi*-simultaneity, rather than perfect coincidence, synchronicity or contemporaneousness. I talk of simultaneity for the sake of simplicity, but my consideration is meant to hold for any two correlated events happening within a restricted time window (i.e., even if just *quasi*-simultaneous).

(a) When did the NCC of conscious experience C (e.g., pain) first appear in the phylogeny of species?
(b) When did the conscious experience C (e.g., pain) first appear in the phylogeny of species?
(c) When does the NCC of conscious experience C (e.g., pain) first appear in the ontogenetic development of a given individual being?
(d) When does the conscious experience C (e.g., pain) first appear in the ontogenetic development of a given individual being?

Questions (a) and (b) are linked to each other, since if the NCC are accurate, the answer to (a) should correspond to the answer to (b) and *vice versa*. Similarly, questions (c) and (d) are linked to each other, since if the NCC are correct, the answer to (c) should correspond to the answer to (d) and *vice versa*.

In addition to these questions, we could ask more general ones concerning consciousness *simpliciter* (in the intransitive sense) rather than consciousness of a specific kind:

(a') When did the physical processes underlying conscious experience *simpliciter* first appear in the phylogeny of species?
(b') When did conscious experience *simpliciter* first appear in the phylogeny of species?
(c') When do the physical processes underlying conscious experience *simpliciter* first appear in the ontogenetic development of a given individual being?
(d') When does conscious experience *simpliciter* first appear in the ontogenetic development of a given individual being?

Again, if we admit the ideal case in which we have a complete understanding of how physical processes are accompanied by conscious experience, the replies to questions (a') and (b') are linked to each other, since the answer to (a') should correspond to the answer to (b') and *vice versa*. Similarly, questions (c') and (d') are linked to each other, since the answer to (c') should correspond to the answer to (d') and *vice versa*.

Answering diachronic questions about consciousness in phylogeny (a, a', b, b') is *de facto* impossible, since it would require having an access to neuroscientific and psychological archaeology that we do not and, realistically speaking, will never have. There are contingent epistemic limitations to what we can know about consciousness in phylogeny[21]. Similarly, answering the diachronic ques-

---

21 See however Feinberg and Mallatt (2016) for a recent attempt to tackle these interesting and difficult questions.

tions about consciousness in ontogeny (c, c', d, d') is at the very least empirically problematic, since it would require having access to accurate subjective reports from newborn babies and young infants – something that we do not and cannot have. Thus, even with respect to these questions we may (and probably will) never have an epistemically objective grip on consciousness in the diachronic framework.

The fact that there are severe epistemic limitations to what we can know about consciousness from a diachronic perspective probably explains why philosophers and scientists alike have not shown much interest in these problems until very recently. The fact that we cannot *know* what has been the case, however, does not entail that there is not a fact of the matter about when consciousness and NCCs appeared in phylogeny and in ontogeny, nor that these questions are not worth asking. Indeed, consciousness, considered as a natural phenomenon with an ontologically subjective mode of existence, has *de facto* a history – very much like any ontologically objective phenomenon –, whether or not this history is epistemically accessible or empirically verifiable.

For a lack of suitable evidence we do not know much about the archaeology of consciousness[22]. We know some things about consciousness as it presently exists – both from our personal phenomenological experience and from what science tell us about it – but not more. Despite our factual ignorance, however, we can easily conceive of different possible (more or less plausible) scenarios describing the history of consciousness. Considering some possible overall historical trajectories of consciousness from the dawn of our planet to the present day is a way of suggesting ways in which the history of this phenomenon might be interrelated with the history of the natural world as a whole. Doing so, in turn, could highlight diachronic questions about consciousness. I proceed in this direction by proposing a thought experiment suggesting what I think is a plausible scenario describing the history of consciousness in the actual natural world.

1.5.1 Consciousness in Time: a Thought Experiment

Consider the universe at the time of the formation of the Earth, approximately 4.54 billions years ago. Imagine our planet in the infinity of space. Imagine the continuous and unstoppable sequence of physical events, the causal chain of events ruled by the laws of nature. Imagine all this activity, all these physical rearrangements of matter and energy, and yet – in the absence of any life

---

[22] On this topic see Feinberg and Mallatt (2016).

form[23] – the absence of any conscious appraisal, of any sort, of what was going on. At a time when life on Earth had not yet emerged there where no terrestrial knowers of any sort[24]. No biology, no psychology, no consciousness. Not only was there nobody home, to misquote Daniel Dennett, but the Earth was no "home" *tout court* – it was just unknown and non appraised matter.

Everything was going on unconsciously (and – philosophically speaking – smoothly) until something, as in every entertaining tale, changed. Sometime between 4.5 and 3.5 billion years ago life on Earth emerged. Gradually, biological systems started to populate the planet. New species evolved. Previously inexistent biological natural phenomena appeared: mitosis, meiosis, photosynthesis, digestion, and so forth. At *some* point during this evolutionary process *some* living beings *somehow* came to have *some* sort of subjective conscious access to the world, a primitive form of consciousness[25].

To make a long story short, time-travel fast forward until 195'000 years ago, when the first members of our species, *Homo Sapiens*, emerged. Assuming that the human neurophysiological architecture has remained fundamentally unaltered since then, we can assume that at this time consciousness already existed. And now take the last leap back to the present. In this very moment you can feel the slight pressure of your feet on the floor. There is something it is like for you to hear the sound of your surroundings. There is something it is like for you to smell and taste drinks or food. Until a moment ago you were consciously mind-wandering and picturing in your mind things that ceased to exist long before you came to life, earth devoid of life, species since long extinct. In other words, there is no doubt that you are conscious and that consciousness in general presently exists[26].

If my sketch is roughly right, consciousness is a feature of the actual natural world now, but it has not always existed. More precisely, consciousness did not exist 4.54 billion years ago because at that time there was nobody on Earth, no living being that could have had any sort of subjective conscious experience, any feeling at all[27]. At that time there was only physical matter, physical properties,

---

[23] For the sake of simplicity here I simply assume that consciousness is a biological phenomenon – a phenomenon that only some living organisms have. I argue for this claim in chapter 3.

[24] Life and consciousness might of course have existed elsewhere in the universe long before that. However, since the point I want to make is that there has been a diachronic transition from a universe without consciousness to a universe with consciousness, the conclusions of the thought experiment do not hinge on this caveat (the timing and location of the transition are not important).

[25] I purposefully highlight the many of "some" to show how little we know.

[26] I take this to be a rather uncontroversial claim. I argue for it in chapter 3 and 5.

[27] See Velmans (2007) for a discussion of discontinuity theories claiming that consciousness

and physical events. This is by no means to say that this was not valuable or interesting as seen retrospectively from our perspective. However no such thing as genuine value or interest existed in such a world precisely because physical matter alone did not consciously matter (I expand on this idea in section 1.5.2).

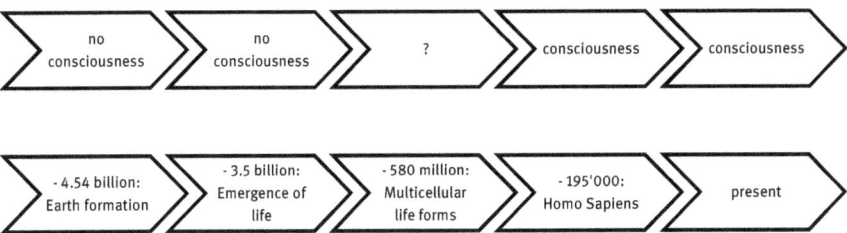

Figure 1

This sketch (Figure 1) is obviously quite rough in form and it surely is controversial in many aspects, but it suffices to highlight a series of questions. Knowing who was the first conscious being, which was the first conscious species, or when, where, and how consciousness arose would be fascinating, but as I have already noted these questions will probably remain unanswered (and, in any case, they are far beyond the scope of the present work). Nevertheless, it seems reasonable to ask what could explain the fact that consciousness presently exists given that admittedly this has not always been the case (i.e., consciousness has not always been there), and that things may have been different. If roughly speaking the above thought experiment is correct, the actual natural world has gone through a diachronic transition from being (i) a world where conscious experiences did not exist, to being (ii) a world where conscious experiences do exist. Notice that this trajectory is not logically necessary. We can conceive a logically possible zombie world Wz having a past history identical to that of the actual world W@ (i.e., no consciousness) but a present history that differs from that of the actual world – i.e., Wz is presently a zombie world, functionally identical to our present world W@, but where conscious experiences do not exist[28]. Why is it that the actual world is presently a non-zombie world, rather than being zombie-like? How can

---

emerged only at a later stage in evolution (such as the one I hold), and continuity theories – arguing that consciousness always accompanies matter. I argue against a form of the latter view, panpsychism, in section 3.2.2.

28 I claim that Wz is logically possible, but I am inclined to think that it is nomologically impossible (i.e., it is not possible given the actual laws of nature). For an introduction to the philosophical notions of zombies and zombie-worlds see Kirk (2015).

we explain the current existence of consciousness in the actual world, given that admittedly consciousness has not always been there and could have disappeared at any time? How come we presently *do* have a conscious access to the world given that, arguably, no such thing had ever existed until a relatively late stage in evolution and it might even have disappeared? How come we have not evolved as mindless organisms in a mindless world, but rather as sentient and feeling beings that *do* mind the world they live in? What happened in the time lapse separating the current state of the world from the time of Earth formation with respect to phenomenal consciousness?

These diachronic questions, unlike synchronic questions such as how to empirically study consciousness and the Hard Problem of consciousness, have been widely overlooked in philosophy and the science of consciousness alike[29]. Taken together, they form what I call the Natural Problem of Consciousness – the problem on which I focus my contribution (see Chapter 3 and following). I am aware that tackling these diachronic questions is a bold and potentially ultimately inconclusive intellectual enterprise. Still, the very contingent fact that consciousness – considered as an ontologically subjective phenomenon – presently exists, justifies raising these questions about *why* it presently exists. We should not dismiss these questions without a thorough attempt to answer them just because it is difficult or impossible to offer verifiable answers. After all, even if we do not end up with definite and uncontroversial answers, there might still be some fruitful philosophical satisfactions along the way.

Before moving to chapter 2, where I introduce the metaphysical mind-body problem and I explain what will be the working assumptions I endorse in the remaining of the dissertation in tackling the Natural Problem of Consciousness, I conclude the chapter with an excursus on the distinction between first-person and third-person perspective and its implications.

1.5.2 Excursus on Objective Knowledge

The term "perspective" is derived from the Latin "perspicere", meaning "inspect, look through", from "per-" (meaning "through") and "specere" (meaning "look at"), which is also the root of the contemporary "perceive". The contemporary meaning of the term "perspective" has not changed significantly. A perspective is a (spatial/temporal/mental) view. The term perspective can be used in different contexts to talk about the range of things that that can be accessed from a given

---

[29] For a notable exception on the scientific side see Feinberg and Mallatt (2016).

standpoint, and is not limited – despite its etymological origin – to the field of vision. Surely, we can talk of perspective in the case of the depiction of depth on a flat surface, but we can also talk allegorically in terms of professional perspectives, and so on. The scope and content of any perspective is intrinsically related to the position and direction or "point of view" of a subject. There cannot be a variation in perspective without a variation of the subjects' point of view.

In some contexts, for example in literature or visual art forms such as films or videogames, it is quite common to draw a distinction between "first-person" and "third-person" perspective. The idea is that a first-person perspective simulates or conveys the mental or visual perspective of somebody (generally the main character), whereas a third-person perspective does not, or does so only indirectly, giving an external and impersonal panorama of the situation. As a literary device, for example, I could write, "I walked towards Rebekka" to induce the reader to identify with me, or I could write, "Michael walked toward Rebekka" to let the reader imagine somebody walking toward somebody else from a free-floating external point of view. Similarly, a first-person visual adaptation of the same story could simulate my point of view when walking toward Rebekka, whereas a third-person adaptation could show both characters from an external perspective, with no precise preference as to which character the spectator should identify with.

I believe that the first-person/third-person distinction, if so intended, is deeply misleading. In the above cases strictly speaking there is no such thing as a third-person perspective at all. Third-person perspective is simply a special kind of first-person perspective, namely a free-floating first-person perspective that is not anchored to a specific character. It is possible to have a perspective from no one (say – no actual character within a story), but *it is not possible to have a perspective from nowhere*. Thus, when in the videogame I see Michael and Rebekka "from the outside", even if I do not see them from a character's perspective, I see them *from* some location somewhere in the (virtual) space around them. This is obvious in this visual case, but it works also in the case of literature, where it is the reader's imagination, rather than an external image, which determines the exact kind of perspective on the action.

Since any point of view has a precise (even if ever changing) location, it is at least potentially somebody's point of view, that is, it is the point of view someone would have if it were there. For this reason I suggest calling any external first-person perspective in the classical sense *transposed-in* first-person perspective (because we are transposed "into" an external character), and any third-person perspective in the classical sense *suspended* first-person perspective (since we are not "into" a specific character, but rather suspended in their world) to highlight the fact that they are not radically different in kind.

There are cases in which the spectator is in a *suspended* first-person perspective (e.g., imagine a shot of a film in which we see two thieves talking to each other in a room) and suddenly there are signs making us understand that we are actually in a *transposed-in* first-person perspective (e.g., the thieves look in camera and point toward the onlooker, the camera shakes abruptly, and an arm holding a gun pops out from a first person perspective making us understand that we are – and have been – seeing the world through a third character's point of view). This, again, shows that the transition from *suspended* (3rd) to *transposed-in* (1st) first-person perspective and *vice versa* is not a radical one. What changes is simply what we take the perspective to signify (e.g., "I am looking from the outside", or "I am looking through the detectives' eyes"), not where the perspective is from.

The same sort of surprise or "breaking of the fictional spell" we have when we discover that we are *transposed-in* another fictional character's point of view is the one we have when the theatre lights up for the pause and we realise we have been watching a film, or when we realize that we have been holding a book in our hands and reading everything. In this case, there is a jump from the thieves' observer's *transposed-in* first-person perspective to what I would call *natural* first-person perspective. The *natural* first-person perspective is our natural subjective perspective. Any sort of experience we have is seen through our *natural* first-person perspective, either opaquely (when we are aware of it being our subjective experience) or transparently (when, say, we are immersed in a fictional world). If you visualise Michael walking toward Rebekka you are seeing them from a *suspended* first-person perspective, sure, but also through your transparent *natural* first-person perspective (since after all it is *you* visualising it). On the other hand, if you see Michael walking toward Rebekka with your own eyes, then you only do so through your *natural* first-person perspective, but not through a *suspended* first-person perspective (since you are seeing them from a first-person perspective that matches your actual location). Even though there are interesting exceptions that could be mentioned and the taxonomy could be embellished further, the details here are not so important. The main take home message here is that we only have experiences from some sort of first-person perspective or point of view – third-person perspective being a special case of first-person perspective.

The reason for this *excursus* is that there is sometimes the bad habit of talking about first-person as indicating something "subjective" and about third-person as indicating something "objective". For example, Searle (1992) sometimes talks about first-person (subjective) ontology and third-person (objective) knowledge. As we have seen, however, any third-person perspective on close analysis turns out to be as subjective as a first-person perspective is. Any perspective is subjective, regardless of whether it matches the point of view of somebody in particular (*transposed-in* first-person perspective / *natural* first-person perspective) or whether it is a non-personified point of view (*suspended* first-person perspective). Thus, it might be questionable if objective knowledge is possible in the absence of subjective points of view (that is, in the absence of an epistemic perspective). Would it make sense to hold that there is a third-person or *objective* epistemology since – arguably – as far as we know, it is only possible to *subjectively* know, think, see or feel?

It makes sense to talk about objectivity to indicate *ontologically* objective objects or properties, that is, mind-independent objects or properties whose existence does not depend on the existence of a subject/observer. However I think that when the epistemic appraisal of these or other properties is concerned access is always and can only be an *epistemically subjective* one. In other words, I believe that there is no such a thing as *epistemically objective* knowledge or appraisal: objectivity exists as a mode of existence, but not as a mode of knowledge. Notice that with this I do not repudiate the distinction between epistemically objective and epistemically subjective *statements* introduced in section 1.2. In fact, a scientific description or theory could be rightfully accounted for as epistemically objective if it refers correctly to true ontologically objective or subjective facts in the world. I simply hold that as long as there is no one *actually conceiving* such epistemically objective descriptions or theories, their semantic/qualitative content – i.e., *what* they express – is latent. A description or theory might be justified and true, but as long as it is not actually meaningful to someone it cannot be considered epistemically objective knowledge.

An example: now you are reading these very black words standing on a white background. Not only you are able to perceive the colour contrast, but by that you also individuate shapes and – more or less consciously – interpret them as forming letters. Those letters put together form words, sentences and propositions that mean something to you. If I write, "imagine the taste of a delicious chocolate cake", you can even try to translate these visible signs that I put on a two-dimensional surface into a mental representation of what it feels like (for you) to eat a delicious chocolate cake. Similarly, if I write "1 + 1 = 2" you can agree that this is an epistemically objective statement. None of this would have been possible 4.54 billion years ago. If this very text had been randomly materialized and lay somewhere on our old "mindless" planet it would have had no meaning, since there would have been no interpreters of these signs. As a consequence the statement "1 + 1 = 2" would have had no *meaning* (even if its content would still have been true). This text, whose meaning is mind-dependent, in the absence of a "translation" from ontologically objective signs to semantic content would have had no meaning and, in this sense, would have carried no knowledge whatsoever.

This applies in a similar manner also to any physical objects or properties that were present on earth 4.54 billion years ago. Even though – say – rocks and water molecules physically existed in a mind-independent way, they lacked all those mind-dependent characteristics that are only expressed in the presence of experience (e.g., that water *feels* wet, and rocks *feel* solid). So, strictly speaking, until there was an epistemic access to the world, water did not have the property of *feeling* wet, and rocks did not have the property of *feeling* solid, nor any other subject-relative phenomenal property, since those properties were not instantiated by any occurrence of water or rocks alone. At most we might retrospectively say that some things had the disposition to have such and such a phenomenal property if there had been conscious beings. But then again having a disposition does not mean exhibiting the phenomenon itself. For example, even if we knew as a fact that water currently has the disposition to appear in the colour "Knifulduc" (a shade of colour invisible to our eyes) to a specific alien life form that does not exist, that does not mean that "appearing Knifulduc" is actually a property of water.

Summarizing, I have suggested that there is no such thing as an *objective* or *third person* epistemic perspective and that everything that can be known can only be known subjectively, from a given perspective. This, in turn, implies that in the absence of beings having a first-person perspective there is no epistemic perspective at all. Now, – as I said in the thought experiment (section 1.5.1) – I assume that before there was life on earth there were also no beings having a first-person perspective. It follows that, if this is correct, before there was life on earth there was no knowledge on earth either. Notice that I am not claiming that there were no ontologically objective facts or things that could have potentially been known (or that had the disposition to be known) in the presence of beings having a first-person perspective (say – if we could go there with a time machine). I am simply claiming that in the *actual* absence of living beings having a first-person perspective (a contingent fact about the history of our planet) all the facts, properties, and things that existed were totally obscure and unknown. Thus, in summary, I am suggesting that the history of our planet has been marked by a gradual transition from the total absence of any epistemic perspective (i.e., in the absence of knowers) to the presence of epistemic perspective(s). This, I argue, is a contingent fact about the history of our planet.

The connection between consciousness and knowledge seems to be a close one. Arguably, long before the development of full-fledged rationality, knowledge, and conceptual understanding, there was already a more primitive form of epistemic perspective in our world: bare phenomenally conscious experience. If this is correct, then by undergoing a diachronic transition from not having ontologically subjective properties to having such properties the world also underwent a parallel transition from not being known to being known. Could it be that the very fact of getting to subjectively know the world is the key to explaining why consciousness presently (still) exists?

2 The Metaphysical Problem of Consciousness

In chapter 1 I have introduced some important general issues concerning the nature, study and ways of thinking about problems regarding consciousness. Before proceeding to the detailed articulation, development, and assessment of the Natural Problem of consciousness, i.e., the problem of knowing why there are presently conscious beings at all (see chapter 3 and following), in the present chapter I consider the metaphysical problem of how consciousness fits in the natural world. First I present the mind-body problem (section 2.1), then I sketch an outline of the major possible metaphysical positions one could endorse (section 2.2), and finally – on this basis – I explain what position I conditionally endorse when tackling the Natural Problem (section 2.3).

The goal of this chapter is to introduce some important notions, highlight some challenges that remain in the background of my research but I *do not* aim nor claim to tackle, and contextualise the working assumptions that I endorse when considering the Natural Problem of Consciousness. Most importantly, I try to eradicate some possible misunderstandings regarding the scope of my research right from the beginning. More precisely, I attempt to prevent the development of some lines of objection (i.e., indirect objections directed at my metaphysical conditional assumptions) that could mistakenly be thought to be potentially harmful to my main thesis. I argue that such lines of objection would inevitably hit the thesis beyond its scope and that therefore they should not be considered as pertinent objections. Even though this chapter is somewhat expository, it is a fundamental step allowing me to introduce the methodological tools and the argumentative strategies I intend to use to tackle the Natural Problem of Consciousness. I aim at making this prerequisite passage as smooth and straightforward as I can without thereby leaving anything important along the way.

2.1 The Mind-Body Problem: the Mind in a Physical World

The sub-currents in the philosophical mind-body literature are numerous, and I follow Chalmers (2002, xi) in classifying them along three major axis of interest: metaphysics, mental content, and consciousness. In this chapter I concentrate on the stream of research focusing on the metaphysics of the mind, tackling questions on the nature of the mind and the relationship between the mental and the physical. The second stream of research, concerned with mental content[30],

---

30 As Markus Wild rightfully pointed out to me, philosophers such as Davidson, Dennett, or

DOI 10.1515/9783110525571-005

tackles questions on the nature of thought, mental representation, intentionality, and the contents of our thoughts, beliefs, desires, and so forth.[31] The third stream of research focuses on questions about the function and nature of consciousness, the main focus of the rest of the present dissertation. The boundaries between these three streams of research are often blurred. The answer to one question depends at least partially on one's take on another – interrelated – question. Nonetheless, this partition of the approaches to the general mind-body problem proves helpful in highlighting what are the chief *explananda* in philosophical research in this field.

The central problem in the metaphysics of the mind, commonly referred to as the mind-body problem, can be summed up by the following all-embracing questions: What is the relationship between the mental and the physical? What is the relationship between mind, brain and body? Much has been written on the topic, with the focus moving along different aspects of the mind-body relationship and considering different aspects of the mind.

2.1.1 The Set of Premises to the Puzzle of the Inconsistent Triad

Regardless of the focus of research one adopts, there is always a problem lurking in the background – a metaphysical puzzle that, in different forms, has been occupying and fascinating philosophers for centuries. A good way to highlight this puzzle is to formulate it as an inconsistent triad – a set of three important principles describing the mind, the physical world, and their relationship that, if held together, yield an inconsistent view[32]. The challenge for the philosopher is to find a way to accommodate these principles in such a way as to formulate a coherent view of the mind, the physical world and their relationship and get rid of the *impasse*. I proceed by presenting individually principle (1) "causal, nomological, and explanatory completeness of the physical domain" (section 2.1.2),

---

Brandom think that the "central" problem is the ascription of mental states, and that the solution to that problem indicates the solution to the metaphysical problem.
31 Intentionality is a crucial trait of most of those states that we label "mental". The paradigmatic examples of intentional mental states are beliefs and desires, that is, mental states which (i) are about, directed towards, or represent something, and (ii) have some meaning or content. I do not consider intentionality in the present work. The *locus classicus* of the discussion of intentionality is Brentano (1874).
32 I borrow this methodological approach from Kim (2005), but I expose the principles with a slightly different formulation. For an introduction based on Kim's puzzle see also Esfeld (2005, pp. 7–17).

principle (2) "physical irreducibility of the mental" (section 2.1.3), principle (3) "causal efficacy of the mental" (section 2.1.4), and by showing why they form an inconsistent triad on the basis of a further guiding principle (4) "the absence of regular overdetermination" (section 2.1.5).

2.1.2 Only a Physicalist Framework Can Account for Causation

Consider the following principle:

> (1) *Causal, nomological, and explanatory completeness of the physical domain:*
>
> For every physical state[33] or physical event P, if at time *t* P has a cause, is submitted to certain laws and allows for a causal explanation, then at *t* P has a *complete physical* cause, is submitted to *complete physical* laws and allows for a *complete physical* causal explanation.

This principle suggests that the physical world is causally complete (i.e., that non-physical causes could not bring about any effects in the physical domain for which there are not also complete physical causes). The key claim of principle (1), more precisely, is that the physical world is complete under causation, that is, that a physical state or event P, if it is caused, is caused by another physical state or event. Principle (1) is a universal philosophical principle (i.e., a principle that applies universally across any physical state or event) resulting from the widely accepted view in contemporary natural science that – since Newton's theory of gravitation – there are universal natural laws (physical, chemical, biological[34], and so on) applying to every object localized in space (that is, as it turns out, physical objects)[35].

Notice that the validity of principle (1) does not necessarily imply a monist materialist ontology, that is, an ontology following which the *only* things that

---

[33] Physical states and events are those states and events that are (or can be) the subject of natural science (physics or special sciences such as chemistry or biology). It is controversial whether psychology is among them.

[34] Markus Wild correctly pointed out to me that there is a debate about whether there are biological laws of nature in evolutionary biology. Still, here the point applies more generally.

[35] For an argument in favour of principle (1) based upon the history of modern physics see Papineau (2002, Appendix). Emergentists such as Carl Gillett deny this principle based on the rejection of the view that there are universal physical laws. However there seems to be no evidence for this claim (e.g., for the claim that in the human brain Newton's law of gravitation is not valid). I thank Michael Esfeld for pointing this out to me.

exist are material entities, states or properties[36]. In fact, the principle says that if a physical state or event P has a cause, it has to have a *physical* cause, but it does not exclude *prima facie* that there can be other kinds of states or events (such as non-physical mental states or mental events), nor that there can be causation processes that do not involve only physical causes and physical effects (such as mental to mental causation, or causal overdetermination).

Any view of the mind-body relationship that does not endorse principle (1) – i.e., any thesis following which a physical state or event P, if it has a cause, does not necessarily have a *complete physical* cause – appears to be in contrast with the widely accepted scientific worldview. In order to reject principle (1), one has to successfully explain *how* a non-physical state or event (say – a mental state M which is not identical or reducible to any physical state) could cause any physical state or event P that has no sufficient complete physical cause. This is commonly known as the *problem of mental causation*. Having to solve this problem is a heavy explanatory burden to carry, and history tells us that those who attempted this route – Descartes *in primis*, with his interactionist substance dualism – did not land on their feet.

Principle (1) is legitimated by what we know about the world from the natural sciences. Denying causal closure, thus, comes down to challenging one of the core principles arising from and guiding our best scientific theories. For this reason, there is a widespread and more or less tacit consensus among contemporary philosophers that (1) deserves to be endorsed and to be considered a non-negotiable condition for a working account of the mind-body relation compatible with natural science.

2.1.3 The Subjective Character of Consciousness is Irreducible

Let us now consider a second principle:

> (2) *Physical irreducibility of the mental (A.K.A. distinctness):*
>
> Some mental states or events have phenomenal properties, and these properties are not identical with, and not reducible to physical properties. At least conscious mental states or events are not identical with, and not reducible to, physical states or events.

This principle is based upon the ontological consideration that physical properties and physically reducible properties are not all there is to reality. In other

---

[36] I thank Michael Esfeld for pointing this out to me.

words, when we think of what kinds of things exist in the world, physical properties and physically reducible properties do not exhaust everything that there is. More precisely, this principle holds that what is missing in a purely physicalist picture of reality is a second kind of property, the so-called *phenomenal* properties[37].

Phenomenally conscious mental states are the paradigmatic kind of mental state that are used to make the point in favour of *distinctness*. Any phenomenally conscious mental state is characterised by the fact of having phenomenal properties. Phenomenal properties are the qualitative properties of subjective conscious experiences such as the "redness" that we perceive when looking at apples, the "painfulness" of pain or the "tastiness" of soups – that is, the properties determining what the experience *feels like* for a subject. Phenomenal properties, according to this view, are not identical with, or reducible to, the properties of any physical state (neither physical properties of the subject, nor physical properties in a wider sense). The reason for this, as we have seen (cf. section 1.2), is that phenomenal properties have a subjective mode of existence, whereas physical properties have an objective mode of existence.

Think for example of pain. Pain resulting from tissue damage is a bodily feeling that has a certain phenomenal quality (or *quale*) for the subject of the pain experience: it *feels* painful. According to principle (2), the "painfulness" of pain is not a physical property, but rather an irreducible phenomenal property. Even if one could give a complete physical explanation of – say – the neural states correlated with (or underlying) that experience in terms of physical properties, one would still have to account for the existence of the qualitative properties of pain experiences (their "painfulness") in *addition* to the former – as highlighted by the Hard Problem (see section 1.4). It seems that if one does not add an account of the phenomenology of pain (i.e., what pain *feels* like) to the purely physical description of what brings about pain at the physical level, one is not telling all there is to know about pain. Indeed, pain without its peculiar *quale*, i.e., the "what it is like" to feel a pain, is hardly a pain at all. If this is true, then this suggests that ontologically subjective phenomenal properties are something distinct from (i.e., not identical and not ontologically reducible to) ontologically objective physical properties.

Notice that this does not at all imply that phenomenal properties cannot be causally or constitutively linked to physical properties in a more or less tight

---

[37] It is an open question whether there are also other properties that are missing in a wholly physicalist picture of reality. Here I focus exclusively on phenomenal properties because this is all I need to highlight the inconsistency of the three principles.

manner[38]. For example, it seems right to hold counterfactuals such as "if humans had no brains, humans would have no conscious states", or "in the absence of the NCC of pain we would not feel pain". In other words, ontological distinctness does not necessarily imply that mental properties could exist independently of physical properties, nor that the phenomenal character is necessarily wholly independent of physical properties. But one thing is to say that brains are causally or constitutively necessary (at least contingently) for the existence of phenomenal experiences; another thing – much stronger – is to claim that phenomenal properties are (ontologically) identical or can be reduced to physical properties of brains, bodies, or matter in general. Distinctness as I intend it only denies the latter – ontological – claim.

In order to counter principle (2) it is not sufficient to claim *that* phenomenal properties are wholly identical with or reducible to physical properties. One would have to successfully explain *how* the peculiar subjective character of phenomenal properties could be ontologically identical with or reducible to physical properties. There have been several attempts to do this, but those I am aware of have a fishy aftertaste[39]. Any solution merging the two kinds of properties seem to deny a mere fact about reality, namely that conscious experience has an ontologically subjective mode of existence (its existence being subject relative), which is distinct from the ontologically objective mode of existence of physical properties (that exist independently of any subject).

Boiling water has several ontologically objective properties (entropy, chemical constitution, and so on) that exist independently of any observer and that can be individuated in an epistemically objective manner. Boiling water can also have – or, better, we can attribute to water – ontologically subjective properties ("*feeling* hot" and "*looking* transparent"), but these properties have a different mode of existence; they exist only *qua* phenomenal properties of a conscious subject. The "feeling hot" property is not identical with, or reducible to, any ontologically objective physical property: neither of water (e.g., heat), nor of the conscious subject (e.g., a brain state), nor a combination of the two.

Consider an auxiliary argument in support of the distinctness claim. According to some people there is something it *feels like* to think, reason and know:

---

[38] I discuss the link between the physical and the mental further when introducing the notion of supervenience (see section 2.2.3).

[39] This is obviously not an argument against identity-theories, but is a good-enough reason to be careful. The *locus classicus* of mind-body identity theory in modern philosophy is Hobbes (1655, chapters 1–6). For a contemporary defence of *a posteriori* psychophysical identity see Bickle (2003).

a phenomenology of thought or cognitive phenomenology[40]. The very act of thinking the proposition P "phenomenal properties are identical with or reducible to physical properties" has a particular subjective phenomenology – it feels in a certain way. This *quale* has an ontologically subjective mode of existence – that is, the phenomenal quality of this thought is accessible only to you, subjectively. Indeed, the phenomenal quality of my thought P might differ from the phenomenal quality of your thought P. If this is right it indicates that, first, the phenomenal quality of the thought is not intrinsically attached to the structure or semantic content of proposition P; it is not the case that every time this proposition P is written down or uttered the same phenomenal quality comes for free with it – regardless of the user. Furthermore, the phenomenal character of any thought P has an ontologically subjective mode of existence, and that is not identical with or reducible to the ontologically objective mode of existence of any physical (brain) properties that is supposedly underlying the formation of specific thoughts. If this is right, this suggests that the very act of arguing for psychophysical reduction or identity – that is, against proposition (2) – can be used as a phenomenological counterargument (i.e., a "performative" self-contradiction) pointing out the former argument as being fallacious. Since when you argue for P there is something it is like for you to do so, and since that has an ontologically subjective mode of existence that is untraceable to the ontologically objective mode of existence of P or to the physical properties of your brain, this suggests that the content of proposition P – "phenomenal properties are identical with or reducible to physical properties" – is false. I take this to support the idea that phenomenal properties are qualitatively *distinct from* and *irreducible to* physical properties.

One may be tempted to dismiss distinctness on the ground that reduction does not imply eliminativism. Indeed, when we causally reduce the surface properties of an object to fundamental physical properties of that object we do so without thereby denying the existence of the surface properties. For example, when we say that water is identical to H_2O we simply redefine a phenomenon and its surface features exclusively in terms of its underlying physical causes. This is not meant to deny the existence of the surface properties of water, but rather to say that all the surface properties – including secondary qualities such as heat or colour and primary qualities such as liquidity – are effects produced by and in principle causally reducible to the "real thing", i.e., the fundamental physical

---

[40] Arguably, we could deny cognitive phenomenology and explain the feeling by association with emotions and moods. Still, my point is that thinking somehow "comes with" subjective phenomenal experience.

properties of H₂O. Saying that water is H₂O, or that heat is the mean kinetic energy of molecule movement, implies that the reality of water and heat do not hinge on subjective experiences (i.e., on how water and heat *appears* to us). This, however, does not come down to saying that there are no subjective *experiences* of water or heat (i.e., that water and heat subjectively *appear* in a certain way) – it just means that we should not define "water" and "heat" in terms of how they *appear* to us, but rather in terms of what they actually are. In Searle's terms:

> "Real" heat is now defined in terms of the kinetic energy of the molecular movements, and the subjective feel of heat that we get when we touch a hot object is now treated as just a subjective appearance caused by heat, as an effect of heat. It is no longer part of real heat. (Searle 1992, p. 119)

For any surface feature such as heat, liquidity, sound, colour, solidity, and so on, we could redefine these phenomena in terms of their underlying physical causes and physical reality, eliminating any *reference* to subjective appearances in the definition. In these cases, reduction does not entail the elimination (i.e., the denial of the existence) of the subjective experiences themselves; it simply excludes them from the definition of the phenomenon.

The problem is that an identification of phenomenal properties with physical properties is not an equally viable option. The reason why we *can* identify water as H₂O, or heat as the mean kinetic energy of molecule movement, is that we can make an appearance-reality distinction; all the surface properties of water and heat can be fully described by the ultimate objective physical reality of these phenomena – regardless of their subjective appearance (i.e., how water and heat subjectively appear to us). As Searle points out, however, we *cannot* make a similar appearance-reality distinction for consciousness. The reason for this is that consciousness consists precisely in the appearances themselves:

> Where appearance is concerned we cannot make the appearance-reality distinction because the appearance is the reality. (Searle 1992, p. 122)

The belief that we can successfully entirely reduce subjective experience to physical reality (or phenomenal properties to physical properties) rests on a fallacy of equivocation. We can reduce heat to the mean kinetic energy of molecule movement because heat does not ultimately consist in how heat subjectively appears. However, where the subjective *appearance* of heat is concerned (i.e., the *feeling* of heat), the subjective appearance does ultimately consist in how heat subjectively appears. Its subjective dimension is essential to the existence of the *feeling* of heat. More succinctly, the *feeling* of heat consists essentially in what heat subjectively *appears* to be. Since in the case of conscious phenomena the reality *consists precisely in* the way in which these phenomena subjectively appear to be, if we were to carve off the subjective appearance and redefine consciousness (e.g., the *feeling* of pain, or the *feeling* of heat) exclusively in terms of its underlying physical reality, we would lose the point of the reduction and of having the concept of consciousness in the first place. In short, we cannot give a fully satisfactory account of what consciousness is without referring to its subjective appearance because the subjective appearance is precisely what makes consciousness what it is.

In the "water is H_2O" case, both the surface properties of water and the fundamental properties of H_2O have an ontologically objective mode of existence: they exist independently from any observer. Water at 20 °C has the surface property of liquidity and has a chemical structure H_2O even if there is no one on earth (the only – epistemic – difference being that no one would *know that* water is liquid). There is no fundamental ontological difference between the reduced properties (liquidity) and the reducing properties (chemical structure), and that is why reduction is possible. In the case of consciousness, however, this does not hold. Consciousness has an ontologically subjective mode of existence (it exists only *for* someone), whereas the physical states underlying it have an ontologically objective mode of existence. The problem of reduction does not arise when trying to reduce ontologically objective properties such as liquidity (or heat) to other ontologically objective properties, but rather when we try to reduce ontologically *subjective* properties such as the *"feeling* of liquidity" to ontologically objective properties, since in this case reduction would come down to the elimination of the subjective dimension.

There are several well-known arguments suggesting that consciousness is not ontologically reducible in the way that ontologically objective phenomena are (see for example Nagel 1974; Kripke 1971, 1980[41]; Frank Jackson 1982[42]). Here is John Searle's version of the argument:

---

[41] Kripke (1980) claims that the expressions on both sides of an identity statement are rigid designators, i.e., they refer to the same object in every possible world. For Kripke, since heat is identical to the mean kinetic energy of molecule movement in every possible world, this identity is a necessary truth. He argues that when we think that the mean kinetic energy of molecule movement might exist in the absence of heat, we are confusing this with thinking that the mean kinetic energy of molecule movement might have existed without being subjectively *felt* as heat, which is indeed possible (Kripke 1980, p. 151). In contrast, Kripke claims that since conscious states could be non-identical with physical states the identity is not a necessary truth and thus physicalism is false.

[42] Jackson (1982) claims that complete knowledge of all the objective facts is not sufficient to know everything there is to know (e.g., knowing all the objective facts about the colour red is not

> What fact in the world corresponds to your true statement, "I am now in pain"? Naively, there seem to be at least two sorts of facts. First and most important, there is the fact that you are now having certain unpleasant conscious sensations, and you are experiencing these sensations from your subjective, first-person point of view. It is these sensations that are constitutive of your present pain. But the pain is also caused by certain underlying neurophysiological processes consisting in large part of patterns of neuron firing in your thalamus and other regions of your brain. Now suppose we tried to reduce the subjective, conscious, first-person sensation of pain to the objective, third-person patterns of neuron firings. Suppose we tried to say the pain is really "nothing but" the patterns of neural firings. Well, if we tried such an ontological reduction, the essential features of the pain would be left out. No description of the third-person, objective, physiological facts would convey the subjective, first-person character of the pain, simply because the first-person features are different from the third person features. (Searle 1992, p. 117)

I find Seale's explanation convincing: if it is clear what is meant by "ontologically subjective mode of existence" and "ontologically objective mode of existence", it is obvious that these are fundamentally distinct modes of existence. A successful ontological psychophysical identification or reduction would require explaining *how* the ontologically subjective mode of existence of qualitative properties of phenomenality could be identical or fully reducible to the ontologically objective mode of existence of physical properties of matter. But how can one successfully bridge the ontological gap without losing anything along the way? I am not claiming that this is impossible, but I can foresee no plausible way in which one could successfully dismiss the distinctness claim.

I take it that ontologically subjective phenomenality is a constituent of reality in the actual world: our world is a world in which phenomenality exists. Furthermore, I take the distinctness of ontologically subjective phenomenal properties and ontologically objective physical properties – due to their different modes of existence – to be a non-negotiable starting point for any reasonable account of the actual natural world. Since denying distinctness requires denying the existence of both an ontologically objective and an ontologically subjective mode of existence, I endorse principle (2). If we agree that a satisfactory account of the actual natural world has to account for everything that exists in such a world, then any such theory has to either prove that distinctness is false, or account for the existence of both ontologically objective and ontologically subjective properties.

---

sufficient to know *what it is like* to see red). This was meant as an argument against *ontological* reduction, despite the fact that it has often been used to raise worries about epistemic reductionism and mistakenly treated as being an epistemic argument.

2.1.4 Mental Causation

Let us proceed to the third principle forming the inconsistent triad:

> (3) *Causal efficacy of the mental (mental causation):*
>
> At least some mental states can have causal efficacy. That is, at least in some cases, the instantiation of mental properties can and does cause other properties, both mental and physical, to be instantiated.

This principle is based upon a folk psychological understanding of how causation works. Following the common sense view, our beliefs, desires, feelings and so on can – and often *do* – have a causal impact on what we think and on what we do. For example, my feeling of thirst can cause a desire to drink (mental – mental causation). This desire in turn, combined with my belief that there is a glass of water on the table, can cause my action of moving towards the table, grabbing the glass, and drinking (mental – physical, or "psychophysical", causation).

This way of seeing ourselves as causal agents (i.e., agents capable of thinking, reasoning, deciding and causally interacting with the physical world) is strongly rooted in our everyday experience and represents a strong intuition about the powers of our minds. This intuition, however, clashes with the problem of knowing *how* mental states can cause other states at all. If you decide to step onto your desk and shout, "I am a free agent!" there is little doubt you will succeed in doing that. Similarly, if – as Searle often says – you decide to raise your arm, in normal circumstances the "damn thing" will go up. However, the problem consists in explaining *how* the mental state that subjectively appears to be the cause of your physical action can cause such a physical action.

There are several reasons for wanting to save principle (3) – the reality of mental causation. According to Kim (2005, pp. 9–10) the most salient one is the need to ensure the possibility of human agency, which – in turn – is a condition for our moral practice. Denying principle (3) seems to reject what makes us morally responsible beings, namely the possibility of voluntarily controlling and influencing (at least partially) the course of events on the basis of our reasons, beliefs, desires, intentions or decisions to act in a certain way. Another reason to save mental causation is that the reality of mental causation seems to be a presupposition for explaining human knowledge and cognition. Reasoning, for example, seems to be a causal process allowing the acquisition of new knowledge and beliefs on the basis of previously existing knowledge and beliefs. Similarly, memory seems to be a causal process of retrieval of information. If mental causation were not real, if – say – it were just an *illusion* of causation, this would strongly discredit the absolute value of agency and cognition, even though this would not necessarily change how we perceive things to be (indeed, "illusory" mental causation is presumably phenomenally similar to "real" mental causation, despite the former lacking any real causal power). One further important reason to save principle (3) is that the reality of mental causation is at the basis of the explanatory practice of cognitive and behavioural sciences. If mental causation were just an illusion, we might have to reassess the explanatory value of the above sciences, despite their apparent success.

In short, the strongest drive for trying to fit mental causation in a theory about the mind-body relationship is a phenomenological one: we experience mental causation to exist. For example, I experience (i) my desire to drink as causing (ii) my action of grabbing a bottle and drinking. If the only way to explain this or any similar experience is to say that the mental state (i) actually caused the physical state (ii), then mental causation has to be preserved. However, if there is an alternative way of accounting for why we *experience* mental causation to exist as we do that does *not* require mental causation to be real (such as in the example of an "illusory" feeling of mental causation mentioned above), I think that there would be no further binding reasons to insist on defending mental causation. For example, if there is a purely *physical* causal story that is sufficient to wholly account both for (ii) my grabbing the bottle and drinking after (i) feeling a desire to drink, and for (iii) my belief that the desire I felt was the cause of the action, then I see no reason why we should postulate a "real" mental causation in addition to that. Many philosophers would not give up this principle with as little resistance as myself, but this does not worry me too much. The only thing that I want to highlight is that, even though no one denies that mental states genuinely *seem* to cause mental and physical states (i.e., even though we *feel* and it *appears to us* that mental causation occurs), if we can explain both the causation process and why the causal process *seems* to us to be a process of mental causation, there is nothing left to explain[43].

2.1.5 The Joint Inconsistency of the Set of Premises

As we have seen, taken separately, there are some reasons to endorse and defend all the three principles – (1) *causal, nomological, and explanatory completeness of the physical domain*, (2) *physical irreducibility of the mental*, and (3) *causal effi-*

---

[43] Notice that this seems to be a case in which we mistakenly base our ontological conclusions (i.e., conclusions about what exists) on the basis of epistemically subjective statements (i.e., what we think is the case). I argued against this sort of processing in section 1.2.

cacy of the mental – on some grounds. However, let us endorse an additional principle guiding the relation of (1), (2) and (3), namely:

(4) *The absence of regular overdetermination:*

If mental states cause physical states, there is not a regular causal overdetermination of these physical states by (both) complete physical causes and additional mental causes.

This principle just says that – say – my action of drinking cannot have both a complete and sufficient physical cause (say, a given neural state) *and* a complete and sufficient mental cause (say, my desire to drink) that are equally causally relevant for the action in question.

In the light of the latter principle, suggesting that there cannot be two causes of a different nature (say, one mental and one physical) both relevant to the same physical effect, it is clear that principles (1), (2) and (3), taken conjointly, lead to contradiction. In other words – if we assume (4) – (1), (2) and (3) cannot all be true without thereby yielding an inconsistent view of the mind-body relationship. In fact, if it is true that (1) the physical world is complete under causation and (2) mental states are distinct from physical states, then it is necessarily false that (3) mental states can cause physical states. On the other hand, if it is true that (1) the physical world is complete under causation and that (3) mental states can cause physical states, then it is necessarily false that (2) mental states are distinct from physical states. Finally, if it is true that (2) mental states are distinct from physical states and (3) mental states can cause physical states, then it is necessarily false that (1) the physical world is complete under causation.

There are different possible ways to get rid of the incongruity. The first one is to deny (4), thereby accepting regular causal overdetermination (i.e., claiming that mental causation genuinely exists, but that for every case of mental causation there is also a complete physical cause explaining the effects)[44]. The second one is to accept (4) and either modify or give up one of the three main principles (1), (2), or (3). The last one is to deny more than one of these three principles, but this would obviously yield a theory that would be even more controversial, since it would refute a wider portion of basic principles about the mind, the body, and their relationship.

In order to successfully meet the challenge of dissolving the joint inconsistency of the three principles, a theory has to provide convincing arguments as to (i) why this or that principle is negotiable or non-negotiable, and (ii) why the con-

---

[44] For such positions see Mellor (1995, pp. 103–105) and Mills (1996). For a defence of (4) see Kim (2003, p. 158).

sequences entailed by the suggested modification of a principle are acceptable for a coherent description of the mind-body relationship. Any major position in the metaphysical mind-body debate can be interpreted as taking some more or less explicit stance with respect to this puzzle, i.e., as suggesting a possible solution to the puzzle of the inconsistent triad. There is no common agreement on any one particular solution to the mind-body problem. This suggests how complex it is to construct a theory about the mind-body relation yielding a balance between theoretical strength and explanatory power that satisfies everyone. Nevertheless, as it should already have become clear, I do have personal preferences among the different positions available. In the next section I sketch the different alternatives in order to highlight my assumptions.

2.2 A Sketch of the Metaphysical Landscape

De facto there is only one way the actual world is. The fact of the matter, for example, is that either the actual world has a monist ontology (see 2.2.1), or it has a dualist ontology (see 2.2.2) – the two being mutually exclusive. Unfortunately, we do not know (and we may never know) *which* way the actual world is. This, however, is only an epistemic problem – a limitation to what we can *know*. Even if there is no way for us to unmistakably find out which of the different options on the metaphysical market is right in describing the actual world, one of these or other possible descriptions is actually right.

The metaphysical landscape comprises different attempts to describe the fundamental nature of reality. To do so, the modal scope of the metaphysical mind-body debate extends beyond the actual world to depict also worlds that are conceivable and possible, but – importantly – not ways the natural world *actually* is. Moreover, the modal scope of the metaphysical problem is larger than life: it includes discussions about the conceivability of zombies, Twin Earth, and other things that almost nobody really believes are actually the case, and that may also be not nomologically possible. I highlight this because, in contrast to this, the Natural Problem of Consciousness is restricted in modal scope to the actual world as it actually is. The metaphysical debate is ignited by an open disagreement about which metaphysical position is the most consistent, unproblematic, and explanatory. This disagreement mirrors another one, often tacit, on the criteria to evaluate the success of any such theory.

In the present section I sketch the main options available on the metaphysical market or – in other words – different ways the actual world could be taken to be. Importantly, my goal is not presenting, analysing and evaluating different versions of philosophical solution-proposals to the puzzle of the inconsistent triad *qua* philosophical solutions to the metaphysical mind-body problem. My goal is not to argue in favour or against one or the other metaphysical position I sketch here on the basis of their internal consistency. I introduce the different options in order to ensure it is clear that the Natural Problem of Consciousness I tackle in chapter 3 arises only (or is salient only) if the actual world fits some particular metaphysical description. In other words, this survey of the metaphysical positions serves mainly as a way to prepare the ground for my claim that *if* we assume that one of these metaphysical pictures (and not the other ones) is correct in describing the actual world, then the Natural Problem arises. This serves as a preventive defence against any assault to the Natural Problem of consciousness based upon an attack of its metaphysical presuppositions. Any such attack, I claim, is not playing the game – it tries to dismiss the problem I want to tackle by targeting its metaphysical underpinnings, something I only *conditionally* assume to be right for the sake of the argument.

2.2.1 Ontological Monism

A first broad group of metaphysical theories about the mind-body relation consists of theories defending one kind or another of ontological monism, the position holding that there is only one fundamental kind of substance or properties forming reality. Any such position denies principle (2) – *distinctness* –, but theories differ as to what is the only fundamental component of reality[45].

On the one hand there is *idealism*, the view holding that reality is fundamentally only mental or immaterial – ontologically subjective –, and that there are no physical things, states, or events – no objective ontology[46]. This view drops principle (1) – *completeness* – *a priori* since it assumes that there are no physical things in reality in the first place and thus the principle of completeness has no reason to be postulated. It rejects (2) since there is nothing but the mental. Finally, it rejects (3) since mental states cannot cause physical states (that do not exist) – although they may cause other mental states. Because of its denial of the three principles and its antithetical role to the basic physicalist framework in

---

[45] I willingly ignore *Neutral monism*, the thesis that both physical properties and conscious mental properties depend upon and derive from a "monist" fundamental level of reality that is neither physical nor mental in itself – thus "neutral". According to neutral monism the intrinsic nature of ultimate reality is neither physical nor mental, but rather between the two (cf. Russell 1919; 1921; 1927a; 1927b; 1956a; 1959; Strawson 1994).

[46] See for example Berkeley (1710; 1713). However, notice that not all versions of idealism *simpliciter* are necessarily immaterialist.

which contemporary philosophy of mind and the sciences work, idealism is nowadays widely considered as an extravagant and unrealistic theoretical possibility rather than as a serious contestant in the race to explain reality.

An opposed monist view is *materialism*[47], the view holding that reality is fundamentally only physical – ontologically objective – and that there are no independent mental properties, states, or events – no subjective ontology. According to materialism, in other terms, all things that exist in the world are bits of matter or aggregates of bits of matter, and there is no thing that is not a material thing. This view holds principle (1), rejects principle (2) since there is nothing but the material, and drops (3) *a priori* since it assumes that there are no mental states in the first place. This view avoids the problems of mental causation by simply denying that the mental exists in the first place.

Among materialist theories we can distinguish Eliminative Physicalism[48] and Reductive Physicalism. Eliminative Physicalism holds that consciousness and phenomenal properties do not exist. Reductive Physicalism, on the other hand, claims that conscious mental states are identical or reducible to more fundamental states – typically physical or neurophysiological states. Following the latter view, having conscious experience *just is* being in a relevant brain state. According to Lycan and Pappas (1972), the difference between a true eliminative materialist and a reductive materialist consists in the fact that the true eliminative materialist claims that common sense mental notions do not pick out anything real and that mental terms are empty (i.e., talking about beliefs, desires, pains and other mental states is like talking about other non-existing things such as centaurs or unicorns), whereas the reductive materialist only claims that mental notions can somehow be reduced to states of the brain.

---

[47] The contemporary successor of materialism is ontological physicalism (or substance physicalism), the view holding that all things that exist are entities recognized by physics, or systems aggregated out of such entities. The notion of physicalism allows to account for "fields" and other concepts widely used in contemporary physics but problematic to fit into the traditional notion of the "material". I keep on using the notion of "materialism" here for the sake of clarity since, at least in its general and unspecified form, "physicalism" affirms the ontological primacy, priority, and basicness of the physical, but does not necessarily imply a monist ontology. Think for example of non-reductive physicalism (or "property dualism"), the claim that even though the physical is ontologically basic and primary, some physical systems exhibit psychological properties that are distinct from, and irreducible to, its physical properties (thus yielding a dualist ontology).

[48] For radical versions of this theory see Wilkes (1984; 1988); Churchland (1981; 1983); Feyerabend (1963a, pp. 49–66; 1963b); Rorty (1965). For eliminativism aimed at qualia see Dennett (1990); Carruthers (2000). For eliminativism about the conscious self or the "Cartesian Theatre" see Dennett (1992), Dennett and Kinsbourne (1992). See also Dennett (1971; 1987; 1991).

The classic formulation of reductive physicalism is the Type identity theory[49], the view holding every type of mental state or process – i.e., every class of relevantly similar instances or occurrences of mental states of processes – is numerically identical[50] to a type of physical or functional state or process of the brain[51]. For example, according to the Type identity theory, the mental state type "pain" is numerically identical – i.e., one and the same thing – with some type of physical state or process (the traditional example erroneously involved C-fibers), and similarly for every type of mental state. The psychophysical type-identity, according to Type identity theory, was to be established *a posteriori* on the basis of scientific investigation and empirical identifications – somewhat like in the identification of type-identities such as "water is H_2O" (a case in which every token of the type "water" has all its properties in common with every token of the type "H_2O"), or "lightning is atmospheric electrical discharge" (a case in which every token of the type "lightning" has all its properties in common with every token of the type "atmospheric electrical discharge"). That is, the idea was that eventually we would have discovered that for every token of a given mental state type (e.g., for every token of pain), such token would have been found to have all its properties in common with every token of a given specific brain state type[52]. Some type identity theorists defended their claim suggesting that if mental properties can be shown to be type-identical with neurophysiological properties there is no problem of causation, and thus no explanatory gap to bridge (Hill and McLaughlin 1998; Papineau 1995).

Type identity theory can however easily be attacked on the grounds that there is a problem of multiple realization. Let us suppose that we claim that pain is type-identical with brain state P1. It might be that indeed in humans pain is type-identical with brain state P1, but perhaps dogs have pain in brain state P2, aliens in galactic plasma physical state P3, and so on. If dogs, aliens, and so on can have pain despite not being in brain state P1, then it is wrong to claim that pain is type-identical with brain state P1. The problem is that if a single mental

---

[49] See Place (1956); Feigl (1958); Smart (1959). The claims of Identity theory can be traced back to Thomas Hobbes (1651; 1655), and Pierre Gassendi (1658).

[50] According to the Principle of Indiscernibility of Identicals by Leibniz, for any x and y, x and y are (numerically) identical if and only if they have exactly the same properties (i.e., if they share the same set of properties). In other terms, if, for every property P, object x has P if and only if object y has P, then x is identical to y.

[51] Notice that the claim that mental state types such as thoughts, beliefs, desires, and so on are identical with physical state types does not commit one to claim the opposite, namely that all physical state types are identical to mental state types.

[52] Notice that NCC research seems to be well equipped to highlight such correlations.

state type (e.g., pain) can be multiply realized by or is identical to different neurophysiological states in different individuals (even just in different individuals of the same species), then such a mental state type cannot be numerically identified with any of its realizers taken individually (see Fodor 1974; Hellman and Thompson 1975). In short, in order to make sense of psychophysical type-identities, the type-identity theorist has to explain *how* mental property types could be identical with neurophysiological property types despite the conflicting evidence (Levine 2001)[53].

Another important theory that deserves to be mentioned is functionalism[54]. Functionalism is not a materialist theory (even if most functionalists are physicalists, claiming that the characteristic causal roles of mental states are realized by physical states of the brain), but I introduce it here because it suggests that the mind is entirely reducible to the functions it performs. Functionalism allows us to avoid the trouble of multiple realization by saying that what makes – say – every heart token belong to the heart type is not the fact that all hearts are constructed in a particular way or of a particular substance, but rather the fact that all hearts have the same function and perform the same sort of causal job. Human hearts, mice hearts and artificial hearts – despite the differences in how they are physically realized – are all hearts only by virtue of the fact that they all play a similar heart-role. Applied to the mind, functionalism suggests that having a token mental state m consists in having a state that does an M-causal job – i.e., that occupies the characteristic functional role that M has within a system. The characteristic functional role of M is determined by the characteristic circumstances connected with the mental state M, namely (i) the standard input circumstances causing the mental state, (ii) the standard output behaviour caused by the mental state, and in some cases (iii) the standard internal connections between M and other mental states. For example, according to functionalism, having the belief that lions roar consists exclusively in having a mental state that is caused by (i) hearing a roar sound (or other typical causes), that causes one to (ii) escape, look

---

[53] Most physicalist theories acknowledge the reality of consciousness, but claim that it supervenes on the physical, it is composed of the physical, or it is realized by the physical, and therefore it is still somehow strictly linked with the physical world. *Token Physicalism*, the theory according to which psychological types such as "pain" are not identical with, or reducible to, a type of physical event, but that each and every individual psychological event-token is reducible to a physical event-token (i.e., each instance of pain is reducible to a physical event), is a handy tool in this sense.

[54] For analytic functionalism see Lewis (1966); for psycho-functionalism see Fodor (1968); for machine functionalism see Putnam (1967).

for a lion, or scream "there's a lion!" (or other typical behavioural reactions) and that causes (iii) fear (or other typically connected mental states).

Functionalism, however, has an important problem in accounting for conscious states. According to functionalism, having a feeling of pain, for example, consists exclusively in having a state that is caused by (i) bodily damage (or other typical causes), that causes one to (ii) cry, go to the hospital, or say "ouch!" (or other typical behavioural reactions), and that causes (iii) a desire for the pain to stop (or other typically connected mental states). However, as we have already seen in section 1.4 when considering the Hard Problem of consciousness, there seems to be more to conscious experience (e.g., the feeling of pain – in this case) than just the function it occupies: pain *hurts*. The trouble with functionalism is that a functional account of conscious mental states does not exhaustively define all its properties – in particular, it does not account for its *phenomenal* properties. The nature of consciousness does not seem to be essentially functional, and therefore it is unlikely that consciousness could be explained adequately in functional terms (Block 1980a; 1980b; Levine 1983, 1993; Chalmers 1996)[55].

2.2.2 Ontological Dualism

Leaving ontological monism and functionalism aside, a second broad group of theories about the mind-body relation consists of theories defending one kind or another of ontological *dualism*, the claim holding that mind and body (or the mental and the physical) are neither ontologically identical nor entirely reducible to each other, but rather coexist in one way or another as two fundamental realms of entities or properties. Any dualist position holds principle (2) – *distinctness* –, but theories differ as to how the two fundamental components of reality – ontologically objective and ontologically subjective – are related to each other.

Cartesian (interactionist) substance dualism is the thesis following which mind (*res cogitans*) and body (*res extensa*) are two fundamentally different kinds of substances[56]. Descartes thus endorsed distinctness (2). The main reason leading Descartes (1637; 1641; 1644) to hold that mind and body are different sub-

---

[55] The central arguments against functionalist accounts of consciousness are based on the possibility of inverted qualia (beings that are functionally identical to us, but experience – say – "redness" instead of "blueness"), or absent qualia (beings that are functionally equivalent to us despite having no qualia at all). Some claim that inverted or absent qualia are not possible (Shoemaker 1981; Dennett 1990; Carruthers 2000).

[56] A substance, roughly speaking, is something that could be the only existing thing in the universe. The attributes of substances are usually called properties.

stances is that the essence of mental substances – i.e., what makes something a mental substance – is the activity of thinking (or of being conscious), whereas the essence of physical substances is physical extension. Descartes famously held that it is conceivable and therefore metaphysically possible for the mind to exist without the body, thereby rejecting the dependence of the mental on the physical. Furthermore, he claimed that despite being separate substances, the mind causally interacts with the physical – thus endorsing (3). Several arguments have been raised against Cartesian substance dualism[57]. Despite some isolated defences[58], there is widespread agreement that these counter arguments are conclusive.

Dualism in its most popular forms nowadays is however not a dualism about substances, but rather a dualism about properties. *Property dualism*, in its general form, holds that mental properties – phenomenal properties in particular – are fundamentally distinct from, not identical, and not reducible to physical properties of matter, bodies, and brains, even though they may both pertain to a single kind of substance, usually physical. This sort of position has several variants.

According to *fundamental property dualism*, conscious mental properties are fundamental constituents of reality on a par with fundamental physical properties (Chalmers 1996), and the existence of conscious mental properties is not ontologically dependent upon, nor derived from, any other properties. This sort of position can lead to *panpsychism* or *panprotopsychism*, the thesis holding that even the most basic constituents of reality have some psychological or proto-psychological properties distinct from their physical properties (Nagel 1979)[59].

Some property dualist theories are sometimes labelled *dual aspect theories* because they take some – but not necessarily all – parts of physical reality (e.g., organisms, brains, neural states or neural processes) to instantiate both physical and mental properties. *Emergent property dualism*, for example, claims that conscious mental properties emerge from complex organisations of fundamen-

---

[57] Most famously, the argument against the possibility of causal interaction between different kinds of substances raised by Elisabeth, princess of Bohemia (see Descartes 1989), and the criticism based on the claim that Descartes made a category mistake raised in Ryle (1949).
[58] For contemporary proponents of new versions of substance dualism see for example Yablo (1990); Swinburne (1986, especially ch.8); Foster (1989; 1991, especially ch. 6 and 7).
[59] I tackle panpsychism in section 3.2.2. Interestingly, panpsychism has been recently advocated by neuroscientists Giulio Tononi (2008) and Christof Koch (2012) on the basis of Tononi's integrated information theory of consciousness (IIT). According to IIT consciousness is identical with integrated information. Since IIT predicts that very simple systems such as photo diodes possesses some limited degree of integrated information, this comes down to saying that photo diodes are somehow conscious.

tal physical constituents. Importantly, according to this view, conscious mental properties are fundamental too in the sense that they are something over and above their physical constituents and are not *a priori* predictable from their physical origins[60].

Property dualism is compatible with *non-reductive physicalism*. For example, one could endorse Token identity theory, saying that for every token mental state, that token is identical with and causally reducible to some brain state token (i.e., a brain state token of any type), and yet claim that mental states types are not identical nor ontologically reducible to brain state types. Non-reductive physicalism holds principles (1) and (2)[61]. According to non-reductive physicalists, conscious properties are realized by underlying neural, physical or functional structures or processes (Kim 1989, 1998). Some have argued against this view that we cannot intelligibly spell out *a priori* how the psychophysical realization occurs (Jackson 2004). However it is possible to hold that an *a posteriori* account of how the psychophysical realization occurs would be explanatorily sufficient (McGinn 1991; Van Gulick 1985).

It is interesting to notice that an important variable distinguishing several versions of dualism consists in the account of the causal relationship between mental and physical properties. *Psychophysical parallelism*, a view famously held by Wilhelm Gottfried Leibniz, is the view that physical states can only cause other physical states, and mental states can only cause other mental states[62]. This view holds principle (2) and the principle of causal completeness (1), but rejects principle (3) since it rejects mental-to-physical causation. *Interactionism*[63] is the view according to which physical states can cause mental states and vice versa. This view, in addition to principle (2), holds the principle of mental causation (3). *Epiphenomenalism* is the thesis according to which mental states exist, but cannot cause anything (neither mental nor physical states). This view holds principles (1) and (2), while it rejects (3) since mental states do not cause anything at all[64].

---

[60] See Kim (1998) for a challenge against emergent property dualism, and Hasker (1999) for a defence.

[61] Both Donald Davidson's Anomalous Monism and John Searle's Biological Naturalism can be described as versions of non-reductive physicalism. In what follows I endorse a version of non-reductive physicalism.

[62] According to Leibniz, it only seems because of God's pre-established harmony that mental states cause physical states or vice versa.

[63] See for example Popper and Eccles (1977).

[64] There is also a fourth possible view following which physical states exist, but cannot cause anything (neither mental nor physical states), but it is rather implausible.

I have not come even close to a satisfactory exposition of the vast array of metaphysical positions regarding consciousness, but so be it. Even if we have left a number of issues aside[65], this sketchy survey of the major possible metaphysical positions one could endorse in the mind-body debate should suffice given the task at hand. My goal here was not expounding in depth all the possible options, but rather sketching the metaphysical background on the backdrop of which the discussion of the Natural Problem of Consciousness arises. In order to conclude the preparatory work and illustrate the assumptions I endorse, in the next section I present the mind-body supervenience thesis.

2.2.3 Supervenience as Framework for Physicalism

A necessary minimal commitment shared by all physicalists (reductive and non-reductive) is Jaegwon Kim's mind-body *supervenience* thesis. There are three slightly different ways of formulating it.

The first version says that beings could not be psychologically different from what they are and yet remain physically identical. In other words, that there can be no mental difference without a physical difference.

Mind-Body Supervenience I:

The mental supervenes on the physical in that things (objects, events, organisms, persons, and so on) that are exactly alike in all physical properties cannot differ with respect to mental properties. That is, physical indiscernibility entails psychological indiscernibility. (Kim 2011, p. 9)

Thus, for example, if nothing changes at the physical level between time $t1$ and $t2$, I cannot undergo a change from feeling pain at $t1$ to not feeling pain at $t2$, or from feeling pain at $t1$ to having a different mental state, – say – hunger, at $t2$. The view that a physical change of some sort is needed in order to have psychological changes is nowadays well accepted in the Philosophy of Mind and a core assump-

---

65 For example, it might be worth mentioning the issue of whether one could combine ontological physicalism with a conceptual non-reductive view holding that the theoretical and conceptual resources that are apt to deal with facts at the level of the underlying substrate or realization level (e.g., physics), may not be adequate to deal also with facts at the realized level, e.g., consciousness (Cf. Putnam 1973; Boyd 1980; Fodor 1974). That is, one may argue that even though chemical, biological, and psychological properties are physically realized, the science of chemistry, biology and psychology provide modes of description that are autonomous from those of physics, and that justify their autonomy.

tion of science. The opposite idea, according to which our mental life could vary despite an invariance at the physical level, postulates an independence of the mind from the physical that has no empirical support whatsoever and that, as a naturalist philosopher, I see no reason to assume.

Notice that the supervenience claim does not imply (nor refute) the reverse thesis holding that beings could not be physically different and yet psychologically identical. Indeed, it seems metaphysically possible that there could be beings that are psychologically indiscernible from us and yet are physically discernible from us – the same mental state being multiply realized by two or more different physical systems. This might or might not actually be the case, but importantly supervenience does not take a stance on whether there can be no physical difference without a psychological difference.

A second formulation of the mind-body supervenience thesis, known as "strong supervenience", suggests that the instantiation of any mental property M in some x is only possible if x instantiates also an appropriate physical property P, a physical base, substrate, neural correlate, or subserving property that acts as supervenience base for the mental property M.

Mind-Body Supervenience II (Strong Supervenience):

The mental supervenes on the physical in that if anything x has a mental property M, there is a physical property P such that x has P, and necessarily any object that has P has M." (Kim 2011, p. 9)

For example, if I have the conscious mental property of being in pain, there is a physical property (e.g. a specific physical substrate, neural correlate, or subserving property) such that I have that physical property, and necessarily anything having such property is in pain. If the mental supervenes on the physical as suggested by the strong supervenience claim, then it also supervenes as suggested by the previous supervenience claim[66].

A third formulation of the supervenience thesis, known as "global supervenience", holds that if there were another world just like the actual one in all physical respects (same particles, atoms and molecules in the same places, and same laws governing their behaviour), the actual world and that world could not differ in any mental respects. In other words, once all the physical "stuff" and the physical laws governing it are in place, the mental "stuff" is in place too.

---

[66] Whether the opposite holds – i.e., whether the first formulation entails strong supervenience – is more problematic. See McLaughlin and Bennett (2014).

Mind-Body Supervenience III (Global Supervenience):

The mental supervenes on the physical in that worlds that are alike in all physical respects are alike in all mental respects as well; in fact, worlds that are physically alike are exactly alike overall." (Kim 2011, p. 10)

More succinctly, as Frank Jackson (1994) puts it, any minimal physical duplicate of this world is a duplicate *simpliciter* of this world.

The supervenience thesis, in all three forms, strictly speaking only affirms that the mental properties of any x co-vary with the physical properties of x. Whenever x instantiates some mental properties, x also instantiates some physical properties. However, the thesis is generally interpreted more strongly, so as to affirm that the mental properties of any x *depend on* or *are determined by* the physical properties of x. This reading of the supervenience thesis captures the common idea that there can be no instantiation of a mental property that is not grounded in some physical property – i.e., that mental properties must have some sort of physical substrate from which they arise. For example, it seems sound to think that the very existence (and maybe also the quality) of George's toothache depends on or is determined by George's physical properties, that is, by his body (specifically, his nervous system), and the way it works. The fact that we are successful in easing George's toothache by intervening on his body (i.e., by altering the physical basis on which supposedly that mental property supervenes) seems to be a good indicator that this idea of dependence is correct[67].

The *Mind-Body dependence thesis* holding that "our psychological character is wholly determined by our physical nature", as Kim puts it (2011, pp. 12–13), and that "what happens in our mental life is wholly dependent on, and determined by, what happens with our bodily processes" (Kim 2005, p. 14), adds on to the "soft" reading of supervenience and explicitly affirms the ontological primacy, or priority, of the physical in relation to the mental – in accordance with folk and scientific assumptions and practices.

It is important to notice that even by endorsing supervenience and the Mind-Body dependence thesis – two key assumptions in a physicalist framework –, there is still enough room to argue in favour of property dualism. In fact, principle (2) – *distinctness* – is wholly compatible with a form of non-reductive physicalism saying that mental states and events are wholly dependent on and determined by our physical (bodily) states and processes – reality being fundamentally formed by

---

[67] Any alternative account would have to explain the consistent covariance of mental and physical properties in some other way, but I do not see how this could be done except by postulating God's pre-established harmony or something along these lines.

physical "stuff" –, and yet that mental properties are ontologically distinct from, and irreducible to physical properties. This is the position I assume in what follows.

2.3 The Metaphysical Background for the Natural Problem

The strategies one may adopt to face the metaphysical puzzle of the inconsistent triad are varied and depend on what one takes to be the core non-negotiable principles on which a good theory of the mind-body relationship has to be construed. I believe that the actual natural world is best described by a metaphysical theory of the mind-body relationship that respects a series of minimal constraints, namely that:
(i) Is consistent and non self-contradictory (*desiderata* of logical validity);
(ii) Acknowledges the existence of ontologically objective reality and facts (*desiderata* of physicalism)
(iii) Acknowledges the existence of ontologically subjective reality and facts (*desiderata* of phenomenological realism);
(iv) Acknowledges a dependence of mental reality on physical reality (*desiderata* of mind-body dependence).

These minimal *desiderata* being met, I believe that the value of any competing metaphysical theories should be measured in terms of a further criterion:
(v) Explanatory richness (*desiderata* of Epistemic Force).

The above *desiderata* suggest what is the skeleton around which – in my opinion – a satisfactory theory of the mind-body relationship could be construed. These guiding principles might of course be contested, but I just posit them here without further justification as a way to guide my choice of assumptions.

In order to satisfy the *desiderata* of physicalism (ii), phenomenological realism (iii), and mind-body dependence (iv), I endorse the principle of distinctness (2), supervenience, and the mind-body dependence thesis. Amongst the different metaphysical positions that I introduced, non-reductive physicalism is compatible with such a framework. Non-reductive physicalism can in fact be a consistent and non-self-contradictory theory (i), it can account for the existence of an ontologically objective reality (ii), it can account for the existence of an ontologically subjective reality (iii), and it can account for a dependence of mental reality on physical reality (iv). It remains to be assessed to what extent non-reductive physicalism is also explanatorily rich (v). Given that non-reductive physicalism is a position that fits my guiding principles, I endorse it conditionally as my metaphysical head quarter.

I have claimed in section 2.2 that depending on what are the principles of the inconsistent triad one endorses as a non-negotiable starting point for a good theory of the mind-body relationship, the task of the philosopher is that of explaining how her position could account for the theses held in the remaining principle(s). A metaphysical solution defending some principles (the non-negotiable ones) without being able to (or without trying to) account for the other strongly held opinions about the mind-body relationship will not be widely accepted as a satisfactory position. For this reason, regardless of the principles one takes as his or her starting point, the *metaphysical* challenge is that of clarifying how that position could accommodate the theses held in the remaining principles. As I have said, I endorse distinctness (2), whereas for the time being I remain agnostic regarding the other principles (causal closure – 1, and mental causation – 3). Given that my purpose is not to tackle the metaphysical mind-body problem, but rather to think about the Natural Problem of Consciousness, I do not have to defend non-reductive physicalism *qua* metaphysical position (nor any other metaphysical position for that matter). The latter is an interesting project, but not one I need to be concerned with for the time being. Thus, in conclusion, I endorse non-reductive physicalism as being true for the sake of argument and, given the conditional assumption of this metaphysical picture of reality, I introduce the Natural Problem of Consciousness.

3 The Natural Problem of Consciousness

The general background required for a clear understanding of the Natural Problem of Consciousness (henceforth "Natural Problem") – i.e., the problem of explaining the present contingent existence of phenomenal consciousness in the actual natural world – is now in place. In the present chapter, after a brief cautionary note, I endorse naturalism (section 3.1), I argue that consciousness is a biological phenomenon (section 3.2), and – finally – I articulate the Natural Problem[68] (section 3.3).

The goal of the discussion that follows is not to build a universal theory (applicable across any possible world) about how consciousness relates to matter, but rather to build a descriptive theory based upon and confined in its modal scope to what appears to contingently be the case in the actual world. In order to be successful this theory has only to be applicable to the actual world, regardless of whether it does or does not apply to any other possible world too.

This is to highlight once more that the Natural Problem is independent from the Metaphysical Problem (even though, as we have seen in chapter 2, it arises only given some metaphysical assumptions). The metaphysical mind-body problem explicated by the puzzle of the inconsistent triad has a modal scope that is larger than life. That is, it considers the mind-body relation across all metaphysically possible worlds, including worlds such as Twin Earth that are conceivable, but not actual. I want to focus on consciousness by restricting the modal scope of my research to the actual natural world only. Thus, I abandon the discussion of mind-body relationships in general (in the totality of possible worlds) and focus instead on trying to describe the current presence of consciousness in the actual natural world. Work on the Natural Problem undoubtedly shares some assumptions with the metaphysical mind-body problem, and some of the problems that we will encounter will undoubtedly match, at least to a certain extent, metaphysical problems. Importantly, however, what I say about the contingent role of phenomenal consciousness in the actual natural world is not committing me to a specific metaphysical claim. At least in principle, someone could agree with my conclusions on the Natural Problem regardless of his or her position regarding my conditional metaphysical assumptions. This of course does not prevent one

---

[68] I call this problem "natural" because it is a problem arising in the context of a version of *biological naturalism* and because in this context it is natural – meaning "spontaneous" – to wonder why consciousness presently exists. Furthermore, by calling this problem "natural", rather than "hard" or "metaphysical", I highlight this problem as being distinct from these other problems of consciousness I sketched in section 1.4 and chapter 2 respectively.

to eventually deploy some of the results of the present work as a supporting argument for this or that metaphysical theory, but that is not my aim.

Introducing the metaphysical problem of the inconsistent triad and its possible solutions in chapter 2 was a methodological strategy. First, introducing the problem of the inconsistent triad makes it much easier to introduce the Natural Problem in the right context. Second, showing the modal scope of the metaphysical problem highlights the fact that the scope of this and the Natural Problem are not coinciding. This is important because it has to be clear that finding a solution to the Natural Problem does neither require nor implicate a solution to the metaphysical problem. Third, by comparing the metaphysical problem with the Natural Problem I highlight the originality of my approach. I think that the project of explaining the contingent presence of phenomenal consciousness in the actual natural world is not easier just because its modal scope is narrower. The Natural Problem is a different, but equally valuable project as the project of answering the metaphysical mind-body problem.

3.1 Adopting Naturalism: A Pragmatic Approach

A "naturalist" philosophical approach has generally speaking two components. *Ontological naturalism* holds, roughly, that reality is exhausted by nature, i.e., that reality does not contain supernatural entities. *Methodological naturalism*, on the other hand, holds that the methods of science are a possible route – if not the only one – that can be used to investigate all areas of reality. The majority of contemporary philosophers accept both ontological naturalism and methodological naturalism so (broadly) characterized[69].

I suggest that we should examine the role of consciousness in the actual natural world within the framework of both ontological and methodological naturalism. In other words, I suggest that we should assume as a starting point that consciousness is part of nature (i.e., it is not a *supernatural* substance or phenomenon), and that science can be used to investigate all areas of reality, including consciousness. I assume methodological naturalism even if it might be debatable whether we are justified in endorsing it.

I think we are justified in endorsing ontological naturalism as a starting point because it is currently the only coherent framework that could accommodate the world as it is described by contemporary science, and there are no viable alternatives in view. Ontological naturalism is probably true. But, regardless of whether

---

[69] cf. the introduction to naturalism by Papineau (2009).

it actually is true, it approaches the truth closely enough to be regarded as a reasonable framework to assume and work with. Given that there are no rational reasons to believe in the existence of supernatural substances or phenomena, and given that the best way to explain nature (as opposed to cultural phenomena) consists in relying upon the epistemic excellence of natural science and its interpretations as a guide, there are good reasons to adopt methodological naturalism and accept the ontological framework natural science works with – ontological naturalism[70].

Adopting naturalism entails important consequences concerning causation. As we have seen, the contemporary received scientific opinion postulates strong restrictions on the kind of entities that can causally influence the physical world (cf. principle of causal closure (1), section 2.1.2). Following the contemporary scientific view, only physical causes – "physical" in the sense of "fundamental natural forces" – can have physical effects. All modern physical theories explain any physical effect on the basis of universal physical properties (e.g. mass and charge) figuring in universal laws. Our view of what these properties are has changed from Newton to the quantum age and will change in the future, but – presumably – the scheme of explaining all physical effects in terms of universal physical properties figuring in universal laws will remain unaltered[71].

It may be contested that there is no guarantee whatsoever that the current scientific account of the range of causes that can have physical effects is the correct one. History after all teaches us that received scientific opinion can shift rather abruptly if a better theory challenges a consolidated one, and that philosophical positions about the range of causes that can have physical effects have been varying accordingly. The last radical change in received scientific opinion, for example, came in the middle of the 20th century with physiological research (especially nerve cell research) giving no indication whatsoever of physical effects which could not be explained in terms of basic physical forces, i.e., forces that occur also outside living bodies. This was interpreted – strongly – as suggesting that *sui generis* mental and vital forces, even if deterministic, could not possibly cause physical effects. Logically speaking the latter conclusion (i.e., the impossibility of *sui generis* mental and vital forces causing physical effects) does not follow from the empirical results; mental forces could after all cause physical effects even if these can equally be explained in terms of basic physical forces. Nonetheless, the received scientific opinion has decisively influenced the devising and the widespread acceptance of the philosophical principle of causal closure

---

70 I thank Markus Wild for suggesting this.
71 I thank Michael Esfeld for his comments and suggestions on this subject matter.

of the physical domain[72]. Perhaps the trust placed on the currently held scientific opinion is misplaced (i.e., perhaps physical effects *can* have non-physical causes), and our philosophical theory is therefore misguided (i.e., the principle of causal closure does not hold in the actual world)[73]. However, there is no positive reason to assume that this is the case, and therefore I think there are *at least* pragmatic reasons to consider current received scientific opinion at least *as if* being true.

We are the children of our time, and we cannot change that. Scientific research progresses by questioning old theories and suggesting new ones, and those theories that have not yet been refuted or substituted are those that we should take as being valid. Accordingly, I think that philosophers who want to say something about consciousness in the actual world ought to proceed on the basis of the assumption that the current scientific assumptions correctly indicate how the world is. Maybe one day these scientific theories will be refuted and the philosophical theories based on them will collapse with them. In that case, there will be a reason to build a philosophical theory on a new basis, but that motivation is not present yet, and there is no certainty that such a change in scientific assumptions will ever occur. Therefore, basing our philosophical research upon the commonly accepted current scientific opinion and framework is the only reasonable and productive way to proceed.

It is easy to see that by adopting ontological naturalism about consciousness – i.e., by claiming that consciousness is a natural phenomenon – we can make room to argue for mental causation even within the framework set by the widely accepted contemporary scientific view. By taking consciousness to be some sort of natural phenomenon we are not excluding it *a priori* from the range of possible causes of physical effects. Given the constraints of contemporary science, if we want to construct a theoretical account of the role of consciousness in the actual natural world that is as compatible as possible with the three key principles about mind, body and their relationship (i.e., completeness, distinctness, and mental causation), buying into ontological naturalism seems to be a promising first step.

It may be claimed that speculating about consciousness and the Natural Problem in a naturalist framework is a scientific job, rather than a philosophical

---

72 Almost no contemporary philosopher challenges this thesis. But see Lowe (2000; 2003).
73 Burge (1993, p. 117), for example, writes: "The flood of projects over the last two decades that attempt to fit mental causation or mental ontology into a "naturalistic picture of the world" strike me as having more in common with political or religious ideology than with a philosophy that maintains perspective on the difference between what is known and what is speculated. Materialism is not established, or even deeply supported, by science".

one. However, as we have seen in chapter 1, natural science deals primarily with ontologically objective facts and has still a long way to go in order to offer a successful explanation of ontologically subjective phenomena such as consciousness. The role of the philosopher consists precisely in suggesting why and to what extent a science of consciousness is possible – for example by claiming that ontologically subjective phenomena are natural phenomena with a special mode of existence – eventually suggesting an approach that the science of consciousness could then develop to verify such a possibility[74].

3.2 Consciousness as Contingent Biological Phenomenon

So far so good. The next step consists in saying something more about the sense in which I take consciousness to be part of nature (i.e., not a supernatural substance or phenomenon). I claim that consciousness, in the actual natural world, is contingently a *biological* natural phenomenon. That is, I claim that consciousness, in the actual natural world, is a phenomenon that is only present in (some) *living* beings.

I have no unquestionable proof that consciousness is only a biological phenomenon, but neither is there an unquestionable proof of the contrary, so there is no better reason to dismiss my hypothesis on this ground. In support of my hypothesis, however, there is at least some direct evidence that some living creatures are conscious, plus some indirect evidence suggesting that some other living beings are probably conscious as well. On the other hand, there is no evidence at all that some non-living beings are actually conscious. As long as it is granted that my hypothesis is plausible, I am quite happy. The hypothesis is falsifiable, but doing so would require providing evidence against it, and I have a hard time imagining what kind of proof would convince me that there are conscious non-living beings. I believe that the plausibility of my hypothesis, coupled with the heavy burden of proof for its dismissal are good enough reasons to accept the hypothesis as a starting point.

The motivation for the claim that consciousness is a biological phenomenon derives from the fact that in everyday practice we ascribe consciousness to some animals (thus, some living beings), but not to non-living things. To begin with, most importantly, the claim is based on the fact that each of us (humans) is a living being, and each of us knows for sure – subjectively, by introspection – that she or he is conscious[75] (i.e., that there is something it feels like to be, for you).

---

74 See also Appendix [4].
75 This idea can be rendered as a modified version of Descartes' cogito: "I feel, I am conscious".

This suggests that being conscious is coextensive at least with being a specific alive human. The claim is backed up by the additional consideration that other members of our species (humans), living beings behaviourally and physiologically very similar to us, also appear to be conscious. We instinctively assume this because there are overwhelming similarities between the way in which you and other people talk about, express, and deal with "pain", "redness", "cold", and other conscious states. Since on the assumption that other humans are conscious (as you know for certain you are) communication is effective and the behaviour is understandable, there are good *prima facie* reasons to believe that being conscious is coextensive at least with being an alive human. It could be objected that other people could actually be only philosophical zombies, functionally identical to you but lacking consciousness, but I think there is no rational reason to suppose that (I dismiss solipsism in section 3.2.1). Over and above this, most people naturally assume that living beings of some other species (e.g., chimps, dogs, cats, and the like) are (phenomenally) conscious too. The reason, again, is justified by the fact that given behavioural reactions of some animals (e.g., a dog's typically "painful" reaction in response to a painful stimulus) are best explained by assuming that those animals are conscious (e.g., feel pain). This leads us to suggest that there are good *prima facie* reasons for believing that being conscious is coextensive with being some sort of alive being (albeit not necessarily *any* sort of alive being). This outcome is negatively reinforced by the widespread agreement that non-living things (e.g., rocks, hammers, computers, robots, and so on) are not and cannot be phenomenally conscious. The final resulting claim, thus, is that consciousness is a phenomenon that – in the actual world – is *only* present in some living beings, i.e., that consciousness is (exclusively) a biological phenomenon.

Of course, one could retort that this classification is based on unreliable folk-psychological criteria, leading to intuitively attractive but possibly groundless inferences from apparent similarities/dissimilarities in behaviour, morphology and so on to similarities/dissimilarities in consciousness. In particular, one could retort that (i) exhibiting conscious behaviour (as some living beings unquestionably do) does not *necessarily* imply being conscious, and – the other way around – that (ii) being conscious does not *necessarily* require exhibiting conscious behaviour (or any behaviour *simpliciter*). Given that the inferences might be unsound, the reasoning would continue, we are not justified in claiming that consciousness in the actual natural world is only a biological phenomenon solely on the basis of the above criteria. The objection is fair, at least on logical grounds. I grant that we cannot know *for sure* who or what is conscious and who or what is not conscious – except for ourselves. However, since we have direct subjective evidence that *we* are conscious but no direct evidence whatsoever that other humans and animals are *not* conscious, or that non-living beings *are* conscious, we still have better reasons to claim that most probably only some beings are conscious, than to argue for the opposing view (i.e., that most probably it is not true that only some living beings are conscious).

In order to resist skeptical arguments, I deploy a pragmatic position. If we want to study consciousness we should be pragmatic and account for the phenomenon as well as we can, relying also – in the absence of better and univocal sources of evidence – on how consciousness appears to be in everyday practice. It might be that appearance is misleading (e.g., that rocks, unbeknownst to us, *are* conscious), but unless there is some evidence for that, I see no reason why we should endorse any such counter-intuitive position to start with, rather than opting for a working hypothesis that accounts for widely accepted folk intuitions about who (or what) can be conscious (e.g., not rocks). There comes a time where one has to face our epistemic limitations (what we cannot know) and accept that if we want to build a sound and constructive theory regarding consciousness in the actual natural world we have to start from the best available – i.e., the most *plausible* – hypothesis, and not stop from speculating just because we cannot know for sure what is the case[76]. The best way to examine the nature of consciousness is to begin with the most plausible working hypothesis available.

On the basis of the available evidence, the most plausible hypothesis is that in the actual natural world (i) only some living beings are conscious. This option is clearly more plausible than claiming that (ii) no living beings are conscious, and/or that (iii) non-living beings are conscious (see section 3.2.2) – since there is no direct evidence whatsoever for those claims. In the absence of support for the latter claims, I believe that it is unfair to criticise the hypothesis that consciousness is a biological phenomenon.

Working in a naturalist framework allows us to construct scientific hypothesis on the basis of probable knowledge, rather than absolute knowledge, and this is all we need here. The available evidence supports the hypothesis that only some living beings are conscious, so this is probably true and – at any rate –

---

[76] A skeptic might claim that there is a possible world in which most humans are zombies and rocks are conscious, and that we cannot know for sure that the actual world is not like such a world. Thus, the skeptic could claim, we should not assume that we live in a world where humans are all conscious and rocks are not conscious, since that may be wrong. I do not want to play the skeptic game. I simply bite the bullet and hold that, if we want to make progress, in the absence of stronger evidence for another view, we should be pragmatic and at least conditionally assume for the sake of argument that – concerning consciousness – the actual world is likely to be approximately as it appears to be – i.e., not a world in which most humans are zombies and rocks are conscious.

the hypothesis can be held as valid unless better evidence is brought against it. Unsurprisingly, the science of consciousness studies the behaviour and neurology of humans and some other animals, but does not study consciousness in non-living things such as rocks or billiard balls[77]. Following methodological naturalism, we can simply follow the lead of natural science and claim that – since this is how science deals with consciousness – it is reasonable to suppose that only some living beings are conscious.

Once one adopts the latter claim as a starting point, it becomes apparent that a promising way to study consciousness is to study it *qua* a natural biological phenomenon. If consciousness is a biological phenomenon, then there are good reasons why we should be able to adopt the methods of the science of biology to study it.

The above reasons suffice to at least motivate the claim that *it is plausible* that consciousness is a biological phenomenon (i.e., that only some living beings are conscious), and thus to justify my conditional assumption of that claim. For the sake of completeness, however, I briefly tackle and dismiss two noteworthy views – solipsism and panpsychism – that could be raised to resist my suggestion.

3.2.1 Dismissing Solipsism in the Actual World

Solipsism, in its general and unspecified form, is the thesis according to which only one's own mind exists. There are different sorts of solipsism, mainly dividable into epistemic and ontological versions.

Epistemological solipsism holds that we can only know for sure about the existence of our own mind. Epistemological solipsism, in other words, does not deny the fact that other people may have mental states or be conscious, but it denies that we can *know* for sure whether anything outside of our mental states exists, since all we can observe is other people's external behaviour. Thus, epistemological solipsism holds that we cannot know whether other people's mental states exist, nor whether they are qualitatively like mine (i.e., whether the inner experiences we have are qualitatively equivalent). I consider that epistemological

---

[77] Interestingly, there are also scientists studying the behaviour and "neurology" of plants. There are no doubts that plants exhibit intelligent behaviour, but it is much more controversial whether they are also conscious in the way some animals are. The botanist Stefano Mancuso claims they are (Mancuso and Viola 2013), but the evidence he presents can also be explained solely in terms of plants detecting complex stimuli and behaving intelligently on that basis – there is no further reason to conclude that plants are conscious.

solipsism ought to be taken seriously, since it makes a reasonable and conservative point about the limits of our knowledge.

Ontological solipsism, on the other hand, says that only one being in the world has mental states – "me". This can lead to the extreme metaphysical thesis following which one's own mind is the only fundamental entity or existing substance (i.e., a monist idealism following which the physical world and other minds do not exist). I believe that ontological solipsism is unwarranted since it proposes an ontological picture of reality according to which I have a mind but nobody else does solely on the basis of epistemological observations (i.e., that I cannot *know* whether others have minds). It is not wise to justify ontological solipsism on the basis of epistemological solipsism because the limits of our knowledge are not necessarily due to a limit of what actually exists; epistemology is not an infallible guide to ontology. More precisely, the fact that we cannot know for sure whether other beings have or do not have minds (or are conscious), is not a sufficient reason to conclude that – ontologically speaking – other beings do not have minds (and are not conscious).

Ontological solipsism is not compatible with the pragmatic naturalist account of consciousness that I suggest because it considers consciousness as a property of a single being – me – (and then again, which me?), rather than as a biological property that some (i.e., many) living beings can have. However, ontological solipsism is an extravagant metaphysical theory that has to be taken for what it is: a metaphysical possibility. Importantly, however, – given the pragmatic account sketched above – there is no reason to take ontological solipsism as a serious contestant to explain how reality (and consciousness in particular) is in the actual natural world.

3.2.2 Dismissing Panpsychism in the Actual World

> [Panpsychism has] the faintly sickening odor of something put together in the metaphysical laboratory. (Nagel 1986, p. 49)

Another noteworthy view besides solipsism that could be raised to resist my claim that consciousness is a biological phenomenon is panpsychism. Panpsychism, in its general form, is the view that the mind is a fundamental and ubiquitous feature of the world, or that – in other words – the physical world is universally psychical or conscious[78]. There are many possible varieties of panpsychism,

---

[78] cf. Seager and Allen-Hermanson 2013; Chalmers 1996; Rosenberg 2004.

but the one that is relevant here would be a form of panpsychism holding that consciousness – despite what we might think – is not confined exclusively to some living (biological) beings, but rather is a phenomenon or property that is widespread at every level of existence, including non-living matter. According to panpsychism of this sort, protons, electrons and quarks are conscious, and so are rocks, planets, and molecules[79].

Panpsychism of this sort is not compatible with the pragmatic naturalist account of consciousness I suggest because it considers consciousness as a ubiquitous property of matter, rather than as a biological property that only some living beings can have. By making consciousness a fundamental and ubiquitous property, panpsychism has the advantage of camouflaging as "less mysterious" the question of how we become conscious. After all, if consciousness is a ubiquitous property, given that we are composed of conscious or proto-conscious atoms, molecules, and cells, there is nothing substantially new or striking in our consciousness; our consciousness is explained as the outcome of the combination of many pre-existing "consciousnesses". However endorsing panpsychism does not come without problems.

First and foremost, I believe that assuming panpsychism about consciousness is unwarranted because this view proposes an ontological picture of reality according to which consciousness is distributed everywhere and at every level of existence even though we have no direct evidence whatsoever for such a claim. There is no reason whatsoever to grant that protons, electrons and quarks are conscious, and that so are rocks, planets, and molecules. In addition to this, panpsychism has its own independent problems, such as the composition problem of explaining how separate conscious experiences had by the micro-sized units (e.g., atoms, molecules, cells) composing our medium-sized body could combine to yield a single unified experience such as the one we usually experience. An interesting related problem for panpsychism[80] is that we have only direct evidence of the existence of consciousness as we experience it, i.e., at the level of medium-sized biologically-defined units (roughly speaking, as body-centred experiences). Assuming that panpsychism is true, why is our experience persistently, reliably and only bound to this level of magnitude, rather than shifting

---

[79] An alternative and weaker version is panprotopsychism, a view holding that the tokens of some fundamental physical type (e.g., quarks) are protoconscious, that is, that they have special properties that are precursors to fully fledged consciousness and that can collectively constitute consciousness in larger systems (Cf. Chalmers 2013; 2016).

[80] I raised this problem in a commentary to David Chalmers at the University of Fribourg on the 15 May 2014. I am grateful to him and the audience for their comments.

also to micro-experiences of the micro-sized physical composing us (e.g., molecules) or to macro-experiences of the macro-sized objects that we participate in physically composing (e.g., cities or planets)? In other words, if consciousness is a more or less evenly distributed fundamental property of matter, given that physically speaking there is nothing special distinguishing a living body from its parts or from other contiguous physical bodies, why would our experience happen to constantly "match" (i.e., achieve and be limited to) a mid-sized level of magnitude corresponding to a biological unit (a body), rather than being smaller or larger than that? I think that in order to be credible, a pan(proto)psychist theory about the actual natural world has to explain *why* we experience exclusively as if from a single and individual "mid-sized" biological unit perspective (our own), rather than exclusively or also from a multitude of micro-unit perspectives (those of the micro units such as atoms composing our body) and/or from a multitude of macro-unit perspectives (those of the macro units such as agglomerations of people or matter of which our body is a component), given that consciousness would be present at every level of magnitude.

Panpsychism, like ontological solipsism, is an extravagant metaphysical theory. However philosophically interesting it may be – given the pragmatic account sketched above – there is no reason to take it as a serious contestant to explain how reality (and consciousness in particular) is in the actual natural world. Given the implausibility of ontological solipsism and panpsychism and – on the contrary – the plausibility of the claim that only some living beings are conscious, there are good reasons to conditionally assume the latter claim as a working hypothesis.

3.3 Why Are There Feeling Beings in the Natural World?

So far in this chapter I have endorsed naturalism (section 3.1), and argued that consciousness is a biological phenomenon (section 3.2). On the basis of this and of the preparatory work executed in chapters 1 and 2, I can now formalize my starting assumptions, show how they lead to the Natural Problem of consciousness, and suggest what needs to be done to articulate a solution to the problem.

Consider the following assumptions compatible with a form of metaphysical non-reductive physicalism:

> *Distinctness principle (2):* Phenomenal consciousness has an ontologically subjective mode of existence distinct and irreducible to the ontologically objective mode of existence of the physical world.

Supervenience and mind-body dependence: Phenomenal consciousness supervenes and depends on neural activity (at least in humans). That is, the current human neural-physiological architecture allows humans to consciously feel instead of merely objectively detecting input stimuli and producing output responses without feeling anything.

Biological naturalism: Phenomenal consciousness is (contingently) a biological phenomenon.

On the basis of the assumptions, for which I have argued, it is possible to formulate the Natural Problem as follows:
1. [Given biological naturalism] At least until the appearance of life in the actual world there were no conscious beings.
2. [Given biological naturalism] Presently, in the actual natural world, there are phenomenally conscious living beings (i.e. consciousness presently exists).
3. *Natural Problem*: What is the best explanation for the current existence of consciousness (i.e., of conscious beings), given that consciousness has not always existed in the history of the actual natural world?

Even though I am proposing the label "Natural Problem", the problem of explaining why consciousness presently exists is not new. This problem is sometimes raised in consciousness literature, even if usually only indirectly and superficially, in the form of underdeveloped comments or notes, or in ways that I find puzzling because of divergence of premises (Dennett 1988; Bringsjord and Noel 1998; Harnad 2010). An exception, however, is the following passage by Polger and Flanagan, developing the problem quite clearly and at length (1995, pp. 2, 5):

> [Consciousness] did emerge in the actual world – possibly among many species. Why? What adaptive significance does being sentient confer that being terrifically intelligent but insentient does not or could not have conferred under the actual conditions of human evolution? [...] If systems "just like us" could exist without consciousness, then why was this ingredient added? Does consciousness do something that couldn't be done without it? – in addition, that is, to bringing experience into the world? [...] Skeptical worries to one side, Homo Sapiens are conscious. Assuming this is true, but that it is also true that there was no metaphysical, logical, or nomic necessity in making us so, why did Mother Nature settle on "being subjects of experience" as a good solution strategy for us, and quite possibly for numerous other mammals and other genera? [...] Why did evolution result in creatures who were more than just informationally sensitive?

The problem raised by Polger and Flanagan in this passage – the fact that it is not clear *why* consciousness did emerge and presently exists in the actual world – is a forerunner of the Natural Problem as I conceive it. I am not however sympathetic with the teleologically-loaded vocabulary they adopt, suggesting that nature is goal-directed and that there somehow *has* to be a reason behind why things exist as they do, a sort of Aristotelian final cause. For example, the metaphorical reference to Mother Nature's "strategic" decisions to act in one way rather than another might suggest that "Mother Nature" acts according to a goal-directed agenda or on a reasonable basis – something that I see no reason to presuppose. This being said I agree with Polger and Flanagan that the Natural Problem is certainly salient, at least in the context of the theory of evolution by natural selection.

Polger and Flanagan suggest that in order to explain *why* evolution resulted in conscious creatures we would have to provide a convincing story about the adaptive advantage of being conscious. In their opinion, however, doing so is very hard and no one has succeeded:

> [There] exist no good stories for why consciousness evolved in this actual world. There are as yet no credible story about why subjects of experience emerged. [...] We might have been zombies. We are not. But it is notoriously difficult to explain why we are not. (Polger and Flanagan 1995, pp. 15, 19)

This difficulty in solving the problem of explaining why we are conscious at all led Polger and Flanagan (1995, p. 13) to call the problem of explaining why consciousness evolved "probably the hardest problem in consciousness studies." This is not reassuring, but there is no reason to panic.

Starting from the same basic intuition of Polger and Flanagan, we can formulate the Natural Problem in the context of evolution by natural selection. This second formulation of the Natural Problem is similar to the first one, but it has the advantage of being more topic-oriented, suggesting the framework in which a suitable answer could be found:

1. [Given supervenience and mind-body dependence] In the actual natural world, a being can be conscious only if geared with a suitably functioning underlying neural-physiological architecture.
2. [Given biological naturalism] Presently, in the actual natural world, there are phenomenally conscious living beings (i.e. consciousness presently exists).
3. [Given biological naturalism + Evolution] In the actual natural world, the current existence of consciousness – indented as biological phenomenon – is the outcome of evolution by natural selection.
4. *Natural Problem:* What is the best explanation for the fact that consciousness has not (yet) been ultimately selected against?

Before moving to the development of a tentative answer to the Natural Problem here outlined, there is some work to be done. First, I need to carefully clarify what

I mean by the expression "being conscious" – something that I have only superficially done so far (chapter 4). Then, I need to support my claim that consciousness presently exists (section 5.1) and that it is subject to a diachronic change (sections 5.2 to 5.5). Finally, I have to introduce the theory of evolution by natural selection and explain how consciousness could enter the picture (chapter 6).

4 Consciousness as Feeling. Defining Criteria

I dedicate the present chapter to introducing terminological conventions I adopt throughout the dissertation. The main goal of this is the clarity of what follows, but I also draw some conceptual distinctions and suggest how we could conceive consciousness. In section 4.1 I introduce and explain why I adopt the notion of intransitive being consciousness, a version of creature consciousness distinguished from state consciousness. In section 4.2 I expose some methodological points related to the definition of consciousness. In section 4.3 I suggest that a being is conscious if and only if there is something it *feels* like to be that being, something it *feels* like for the being (i.e., if it is phenomenally conscious). I conclude with a digression on the distinction between consciousness and self-consciousness and their relation to each other (section 4.4).

An important reason for endorsing the terminological conventions I propose rather than other ones is that the former are simple enough to be easily intelligible and applicable by anyone – including non-philosophers – thereby permitting a clear understanding and discussion of the overall picture of consciousness that I want to consider. This does not mean that alternative terminological conventions are less valuable. However, the ones I am proposing are particularly apt for the task at hand. I think that reliance upon unnecessarily convoluted philosophical jargon would obtrude the way rather than make things clearer.

4.1 State, Creature, and Being Consciousness

In philosophical consciousness literature, the praxis is to talk either in terms of "state consciousness" or in terms of "creature consciousness" (cf. Rosenthal 1986). There is a philosophical agreement on that:

> [We] should distinguish *creature* consciousness from *mental-state* consciousness. It is one thing to say *of an individual person or organism* that it is conscious (either in general or of something in particular); and it is quite another thing to say *of one of the mental states* of a creature that it is conscious. (Carruthers 2011, section 1, original emphasis)

The Natural Problem raises a question about "creature consciousness." In order to explain why I think this is the case, I begin by presenting some terminological suggestions helping to distinguish state consciousness from creature consciousness. In section 4.1.1 I present the question of state consciousness, questioning this approach. Then I present the notion of creature consciousness (4.1.2), and a variant thereof (section 4.1.3). I proceed to introduce the notion of intentionality and transitivity, explaining that I am interested in intransitive creature consciousness (section 4.1.4). I conclude with an excursus on the debate between first-order and higher-order theories of consciousness (section 4.1.5).

According to the ordinary philosophical use of "mental state", a being can have several mental states at once[81]. That is, for example, you can have both a thought and a perception – two different and independent mental states – at the same time[82]. Thus, speaking in terms of "state consciousness", it is possible to say that a being has simultaneously a *conscious* thought and an *unconscious* perception, or vice versa. The same does not apply to "creature consciousness", since saying that a being is both "creature conscious" and "creature unconscious" at the same time would be a contradiction.

In order to spell out the difference between "creature consciousness" and "state consciousness", I recommend the use of the expression "global state of mind" as counterpart to the notion of "mental state(s)". Following the tradition, I call "mental state" any specific psychological state of an individual being (or "creature") at a certain time (e.g., *a* thought, *a* perception, *a* belief, and so on). On the other hand, I call "global state of mind" the global psychological state of an individual being at a certain time. According to this definition, at any given time an individual being is in *some* – but only *one* – global state of mind. This global state of mind may, but does not have to, be spelled out in terms of the various particular mental states that compose it[83]. This comes in as a very handy terminological distinction to talk about consciousness because it allows us to distinguish "state consciousness" claims about particular mental states (e.g., *a* desire) of an individual being, from "creature consciousness" claims about the global state of mind of an individual being.

---

81 I propose to use "being" as a general term X that can be adopted to portray each and every bearer of consciousness intended as a property. More in detail, I propose to use "being" as a general term X such that only beings (Xs) can be conscious and there are no conscious Xs that are not beings. The term "being" is loose enough to ensure that no potentially conscious X is excluded *a priori* by the use of this term to indicate the kind of Xs that can be conscious. I expose the reasons for this terminological suggestion in sections 4.1.2 and 4.1.3.
82 I have had an explicit confirmation of this in a discussion with David Rosenthal in the context of his postgraduate seminar in the spring 2015 at CUNY Graduate Center. The idea of opposing "mental state" with "state of mind" has arisen in this context.
83 Indeed, it is an open question whether the global state of mind is merely the composition of individual mental states. The linguistic taxonomy is not a reason to draw robust boundaries and suppose that – say – audition and olfaction are natural kinds. I thank Karen Neander for pointing out this to me.

Following the tradition I claim that, at any time *t*, an individual being has a *phenomenally conscious* mental state if there is something it is like to be *in that particular mental state* for the being. Conversely, I claim that, at any time *t*, an individual being has an unconscious mental state if there is nothing it is like to be *in that particular mental state* for the being. In addition to this, I suggest that, at any time *t*, an individual being is in a *globally phenomenally conscious state of mind* if there is *something it is like to be for the being* (i.e., if the being *feels simpliciter*). Conversely, I suggest that, at any time *t*, an individual being is in an *unconscious state of mind* if there is nothing *it is like to be for the being* (i.e., if the being does not *feel simpliciter*). It follows from this that a being in a globally unconscious state of mind cannot have any conscious mental state (even though it may have several unconscious mental states), whereas a being in a globally conscious state of mind (i.e., conscious *simpliciter*) necessarily has at least one conscious mental state (regardless of whether it also has unconscious mental states).

I reserve the term "feeling" exclusively to refer to any global phenomenally conscious experience that an individual being has at a given moment in time when in a phenomenally conscious state of mind. That is, when I talk about "a feeling" I never refer to a particular mental state; rather, I refer to a globally conscious state of mind. Every feeling has some sort of phenomenal character (global quale), and different feelings differ in terms of their specific global phenomenal character or quality (global qualia) – that is, *how* they feel like *for* the being[84].

4.1.1 State Consciousness

It is common practice amongst philosophers working on consciousness to investigate the notion of "state consciousness", focusing on what it means for particular mental states to be conscious (cf. Rosenthal 1986, 1990; Gennaro 1995; Carruthers 2000). The reason to talk in these terms is that it is generally agreed that consciousness can, but need not be a property of particular mental states. Since it is generally agreed that mental states such as thoughts, beliefs, desires, and so on, can be conscious, but can also be unconscious[85], the central question that philos-

---

[84] More about this in section 4.3.4.
[85] I think it is uncontroversial that particular mental states of a conscious being may not all be conscious. The opposite claim, that being conscious is part of what it is for a state to be a mental state or – in other words – that consciousness is the essence of mentality and that we can explain mentality in terms of consciousness, was central to Descartes. See Armstrong (1978) for an anti-Cartesian account of consciousness.

ophers working on "state consciousness" raise is what it is for a mental state to be conscious as opposed to being unconscious:

> Assuming that not all mental states are conscious, we want to know how the conscious ones differ from those which are not. (Rosenthal 1990, p. 729)

There are two options to explain by virtue of *what* conscious mental states are conscious. As Dretske (1997, p. 7) explains, according to higher-order theories of consciousness, a mental state is conscious by virtue of its being an *object* of creature consciousness. A conscious mental state of a being or subject S is an *object* of creature consciousness "by S being conscious of it" – i.e., *of* that mental state. According to first-order theories, on the other hand, a mental state is conscious by virtue of its being an *act* of creature consciousness. A conscious mental state of a being or subject S is an *act* of creature consciousness consisting in "S being made aware (so to speak) *with* [the conscious mental state] – by its occurrence in S making (i.e., constituting) S's awareness of something". Discussing what makes a particular mental state conscious is an interesting philosophical question, but it is not crucial here[86]. The Natural Problem in fact does not consist in explaining what makes a particular mental state conscious, but rather in explaining the present contingent existence of consciousness *simpliciter* (in an intransitive sense), i.e., why there are presently beings that are "creature conscious". Thus, the terminology of "state consciousness" is not appropriate to deal with the task at hand.

In addition to this, even though talking about particular mental states being conscious might be philosophically interesting, I think that the very idea of talking about consciousness in the context of particular mental states (unlike in the context of states of mind) is also potentially misleading.

First, on a superficial reading, the expression "conscious mental states" might be erroneously taken to suggest that mental states are (or can be) themselves conscious. But surely mental states are not and cannot be conscious, at least not in the way I will shortly define it: there is nothing it feels like to be a mental state (belief, desire, ...) *for* the mental state itself! This is of course not what is denoted by the expression "conscious mental states". All mental states are states of some being or subject S, and when we say that a mental state M is conscious we only mean that M has the property of being conscious by virtue of the fact that S is conscious of M (according to higher-order theories) or that

---

[86] For the sake of completeness I briefly consider the debate between higher-order and first-order theories of consciousness as excursus in section 4.1.5.

S has a mental state M (according to first-order theories)[87]. For example, saying that Martha has a conscious mental state – say – a conscious visual perception of a glass, comes down to saying that *Martha* is conscious of the glass, *not* that the visual perception itself is conscious of the glass. Conversely, saying that Martha has a non-conscious visual perception of the car passing by comes down to saying that Martha is not conscious of the car, *not* that the visual perception itself is not conscious of the car. This suggests that even though the expression "conscious mental state" might be handy, it does not make things more intelligible – on the contrary, it may incorrectly suggest that some mental states have the property of being conscious when actually consciousness is ultimately a property of beings.

A connected reason why I think that the expression "conscious mental state" is potentially misleading is that it seems to presuppose, and it may suggest, some sort of realism about the isolated existence of particular mental states. In fact, ascribing consciousness to particular mental states may suggest that particular beliefs, desires, and so on, are clearly distinct from each other (i.e., they exist in isolation, even if they may be connected), and exist independently of whether or not they are consciously perceived by the being in question. For example, according to the standard definition of mental states, we may say that Arthur has simultaneously a visual perception of a cake, a belief that the cake is good, and a desire for cake. The perception, the belief and the desire are considered as isolated, even if connected, particular mental states of Arthur. Only if this is the case does it make sense sense to attribute consciousness individually to all of them, for example by saying that some mental states (say, the visual perception) are conscious, whereas others (say, the belief that the cake is good and the desire) are unconscious. Talking about conscious and unconscious particular mental states, in sum, might suggest that *de facto* we have clearly identifiable mental states

---

[87] It is not unusual to find expressions such as "being S is conscious" equated with "being S has conscious mental states". Following this interpretation, a conscious being would be defined as a being having conscious mental states. Unfortunately, this definition is non-explanatory and rather pointless unless one can first give a clear account of what conscious mental states are and how they differ from unconscious mental states. But the distinction between conscious and unconscious mental states is precisely that the former are "conscious" whereas the latter are not. Thus, in order to explain what conscious mental states are, one has to explain what the adjective "conscious" means in the first place, but one clearly cannot do that by defining it as "having conscious states", the price to pay being the circularity of the explanation. It follows that explaining the meaning of "S is conscious" as "S has conscious mental states" is unsatisfactory, since it just moves the worry to the state consciousness level and back, without telling us anything worth knowing about what "conscious" truly means.

such as visual perceptions, beliefs, desires, and so on, and that consciousness is a property of such individual mental states.

Suggesting that consciousness applies individually to particular clear-cut mental states (e.g., that "a visual perception" or "a belief" is conscious) is at the least phenomenologically odd. Consciousness does not (at least, not usually) manifest itself as a property applying limitedly to a particular and well-defined mental state taken in isolation (e.g., a particular perception, belief, desire, and so on). Phenomenologically, I never have the impression of being conscious in such a directed and clearly structured manner – it does not feel this way. Consciousness presents itself phenomenologically as a much more complex, holistic and somewhat fuzzy property encompassing different sensory modalities and different sorts of mental states all at once. Of course, if someone asks you to tell whether you are visually conscious of a specific target ✚ – as is often the case in scientific consciousness research – you might or might not reply positively. But saying that you are (or are not) conscious of that target is just a way of conveying the idea that ✚ is part (or is not part) of your globally conscious state of mind! That is, saying that you are conscious of that target does not mean that the only thing that you are conscious of is ✚, nor that your conscious experience of ✚ is clearly phenomenologically carved off from any other conscious mental state you might have at the same time. Since considering consciousness as applying to particular mental states seems to presuppose and suggest a phenomenologically weird and – in my opinion – unjustified view of what conscious experience is like, I think that talking about consciousness in these terms is a bad philosophical habit that would better be abandoned. I think it is sounder to describe consciousness as a property of a global state of mind of an individual (or – more simply – as a property of beings), rather than describing it as a property of particular mental states[88].

I think that the main reason why the "state consciousness" vocabulary is so popular and widespread is that it is the best suited, for contingent pragmatic reasons, to be adopted in scientific consciousness research. In the neuroscience of consciousness, as we have seen in section 1.3, it is common practice to limit the focus of research to the individuation of NCCs of the contents of consciousness. Empirical research of this sort, in order to effectively trace correlations for kinds of content, disregards the complex, holistic and fuzzy nature of conscious experience and focuses exclusively on how consciousness transitively applies to

---

88 Notice that we can further investigate *what* a being is conscious *of* (what is the phenomenal content of his globally conscious state of mind) even if we abandon this sort of realism about particular mental states.

specific targets. Experimenters do not try to find out what is correlated with consciousness *simpliciter* or with a subject's globally conscious state of mind; they try to find out what correlates with consciousness *of* a particular kind of target in a perceptual task. They examine what distinguishes at the neural level the set of trials in which a subject S reports being conscious *of* the target (e.g., a visual or auditory target) from the set of trials in which a subject S reports not being conscious *of* the target, and on this basis they suggest what kind of neural activity seems to be involved when we are conscious of this or that. I say that the "state consciousness" vocabulary suits this sort of research because what NCC research does is contrasting cases in which a particular (artificially) isolated mental state (i.e., a perception of something) is conscious, with other cases in which the same mental state in not conscious.

Once we realize what are the severe artificial experimental setup conditions imposed by empirical consciousness studies (I am thinking in particular of NCC research), it becomes clear why conscious experience is described as applying to clearly partitioned and independent particular mental states. The fact that in consciousness studies, for technical, methodological and pragmatic reasons, it is handy to restrict our analysis on whether a being is conscious *of* a particular target – however – does not imply that "state consciousness" is a good characterization of how consciousness is in natural settings. Thus, to conclude, even if the subjective reports in empirical studies might be taken literally as suggesting that consciousness applies to particular clear-cut mental states, it seems that this interpretation is unwarranted.

Given that talking about consciousness as a property of particular mental states is potentially misleading, I confine myself to talking about conscious or non-conscious global states of mind of given beings.

4.1.2 Creature Consciousness

Instead of ascribing consciousness to some mental states and trying to understand how conscious mental states differ from unconscious mental states, it is possible to ascribe consciousness to creatures. Creature consciousness comes in two variants: an intransitive and a transitive one. A creature can be described as conscious *simpliciter* (in the intransitive sense), or as conscious *of* a specific target (in the transitive sense). I do not focus on the intentional dimension (i.e., the aboutness) of transitive creature consciousness[89] because, as already explained,

---

[89] See section 4.1.4 for some considerations about intentionality.

for the Natural Problem it is sufficient to cast the question in terms of intransitive creature consciousness: why are there beings that are conscious *simpliciter*?

It is usually agreed that as far as intransitive creature consciousness is concerned there are no deep philosophical difficulties specific to the topic of consciousness (cf. Carruthers 2011):

> There does not seem to be anything especially philosophically problematic about intransitive creature-consciousness as such. [...] And in so far as there *is* anything problematic about this form of consciousness, the problems derive from its putative conceptual connections with *state-consciousness* (Carruthers 2000, p. 10, original emphasis).

Rosenthal (1990, pp. 729 ff.) formulates the question of intransitive creature consciousness as follows:

> [...] what it is for a person or other creature to be conscious, that is, how conscious creatures differ from those which are not conscious.

According to Rosenthal the contrast between being conscious *simpliciter* and not being conscious *simpliciter* is "reasonably transparent", and there is no "special mystery" about what intransitive creature consciousness is. In his view, to be conscious in the intransitive sense a creature must be awake and sentient (1990, p. 730). Even though I am not convinced that everything is as easy as Rosenthal suggests[90], I think that adopting the framework of intransitive creature consciousness is a good way of approaching the Natural Problem.

This being said, I find the expression "*creature* consciousness" unsatisfactory. Adopting expressions such as "conscious creature", "conscious organism" or "conscious animal" may have the undesirable effect of tacitly suggesting that whatever is not incorporated in the definition of "creatures", "organisms", or "animals" is *a priori* not a viable candidate to be conscious. It may or may not be that animals can be conscious whereas robots or trees (that is, non-animals) cannot be conscious, but we do not know for sure which kinds of beings are or can in principle be conscious: it may be that there is something that is not a creature, an organism or an animal, and yet is or can be conscious. Since we do not know what is actually the case, I think it would be better to avoid expressions such as "conscious animal" or "conscious organisms" that, because of their high specificity, might suggest that there are *a priori* reasons to think that only animals or organisms are conscious and that whatever is not an animal or an organism cannot be conscious. For the same reason, I think it would be better to avoid the

---

[90] I think these conditions are not sufficient. I come back to this later.

expression "creature consciousness" given that it could be taken to suggest *a priori* that only creatures – that is, literally, beings that have been created (whatever that means) – can be conscious.

4.1.3 Being Consciousness

As an alternative to the expression "creature consciousness" I propose the all-inclusive expression "being consciousness". In what follows, I assume that only beings can be conscious, i.e., that there is nothing that is conscious and is not a being of some kind. I adopt the substantive "being" rather than "creature", "organism", "animal", "entity", "thing", "individual", "system" or similar expressions because I want my terminology to remain neutral, at least at first, as to which kind of being (natural or artificial, organic or inorganic, animal or non-animal, divisible or indivisible, simple or composed, and so on) could be conscious. The term "being" indicates that the being in question has to exist (or quite literally "be") – that is, it has to have a *physical* existence, but adds no further commitment regarding the *kind* of physical being that it has to be. By talking about conscious or non-conscious beings rather than about conscious or non-conscious creatures, organisms or animals one can refer to whatever kind of being could be conscious, regardless of what it is and of how it is individuated. Adopting this terminology allows one to talk about consciousness neutrally, without committing *a priori* to the unverifiable claim that only specific kinds of beings can be conscious and excluding only the possibility of the existence of conscious non-beings (such as conscious mental states). Notice that this is fully compatible with my assumption that we have good reasons to suppose that in the actual natural world consciousness is *contingently* a biological phenomenon, i.e., a phenomenon that only some living beings *actually* have.

4.1.4 Intentionality and Transitivity

I have said (section 4.1.2) that a creature – or better, a being – can be described as conscious *simpliciter* (in the intransitive sense), or as conscious *of* a specific target (in the transitive sense). Even if I do not focus on the intentional dimension (i.e., the aboutness) of transitive creature consciousness, I still want to say something more about intentionality in general and as applied to "being consciousness".

In philosophy, mental states (e.g., beliefs, perceptions, and so on) are often taken to be about or directed towards specific objects or states. For example, beliefs are beliefs about something (i.e., they are not beliefs *tout court*), and one's visual perception is usually a perception of something (i.e., it is not a perception *tout court*). This "aboutness" feature of mental states is known as *intentionality*. Accordingly, the content of intentional mental states is called *intentional* content. Within the scope of "state consciousness" it is reasonable to claim that transitive conscious mental states such as being conscious *of* a given target are intentional and have an intentional content. Being conscious *of* the knock on the window or *of* the coffee aftertaste are intentional mental states because they are conscious states that are about – *of* – , or refer to, an intentional object or a state; in this case, the knock and the coffee aftertaste. Even though there are cases (such as consciousness *of* a pain) where it is controversial whether that counts as intentionality or not, as a rule of thumb within the scope of "state consciousness" conscious mental states are taken to be intentional.

Within the scope of " being consciousness" it is possible to consider both the intransitive dimension (e.g., a being S is conscious *simpliciter* – i.e., she is in a – some – conscious state of mind) as well as the transitive dimension (e.g., a being S is globally conscious *of* X – i.e., her globally conscious state of mind is about X). In the latter scenario, where "being consciousness" is interpreted transitively, consciousness is considered an intentional state of mind. The transitive use of "being consciousness" specifies the intentional content of the global conscious experience of a being X. Instead of just saying (intransitively) "Sabrina is conscious" – i.e., conscious *simpliciter* –, it further specifies transitively *what* Sabrina is globally conscious *of*. Doing so might, but does not have to, require endorsing the terminology of "state consciousness". In fact, we might say that Sabrina is globally conscious *of* the conjunction of the pizza on the table, the Beatles song playing in the restaurant, as well as plenty of other things (or something of the sort) – without thereby committing to the claim that the intentional content of the global experience is the result of the sum of the separate intentional contents of particular individual conscious mental states composing the conscious state of mind (e.g., consciousness *of* the pizza + consciousness *of* the Beatles song, and so on). The transitive use of "being consciousness" could thus be very specific in describing the global intentional content of a being's global conscious experience even without committing to a "state consciousness" vocabulary. Being transitively conscious *of* something and being intransitively "being conscious" *simpliciter* are interconnected. A being that is conscious *simpliciter* has to be conscious *of* something, but does not have to be conscious *of* everything. A being that is unconscious *simpliciter*, on the other hand, is a being that is not conscious *of* anything at all.

The issues of intentionality and intentional content are very interesting, but too wide and complex to be considered in depth in the present work. The Natural

Problem only requires explaining why there are beings that are conscious *simpliciter*, thus I focus on why intransitive "being consciousness" presently exists.

4.1.5 Higher-Order Theories of Consciousness

Higher-order theories of consciousness – the contender of "first order theories" in the debate regarding "state consciousness" (see section 4.1.1.) – propose an account of what consciousness is that is radically different from what I have in mind. In what follows I briefly explain why I do not find higher-order theories of consciousness compelling, claiming I do not consider them for the Natural Problem.

Particular mental states can, but need not be conscious[91]. Higher-order theorists, working in the "state consciousness" framework, claim that the difference between conscious mental states and unconscious mental states consists in the fact that conscious mental states are mental states of which being S is aware (i.e., mental states of which the subject is transitively creature-conscious), whereas unconscious mental states are mental states of which being S is not aware (i.e., mental states of which the subject is not transitively creature-conscious). According to this view, for example, a particular mental state (e.g., my desire of chocolate) is conscious if and only if I am aware of the desire, and it is unconscious if and only if I am not aware of the desire. Following this view, conscious mental states are states that can be defined as the objects of a higher-order representation of some kind – a higher-order perception, experience, thought or belief[92].

Higher-order theories of consciousness are comprehensive theories of consciousness trying to explain the distinctive properties of consciousness in terms of the relation between a conscious state and a higher-order representation of

---

[91] The claim that there are conscious and unconscious thoughts, beliefs and desires is rather uncontroversial. However it is unclear whether some other kinds of mental states – typically, perceptual states (e.g. visual perception) – really could be unconscious. Higher-order theorists claim that there is unconscious visual perception. A usual way to suggest this idea consists in the example of absent-minded driving (Armstrong 1968). There is a wide amount of empirical data that is usually taken to support the claim. However, first-order theorists such as Dretske (1995) claim that the data are problematic because visual perceptual states are not unconscious in the relevant sense. The important distinction here is the one between "access consciousness" and "phenomenal consciousness" (Block 1995). According to Block even though visual perception may not be access-conscious at all times, it is always phenomenally conscious (see also Tye 1995; Nelkin 1996).

[92] This is Lycan's "simple argument" for a higher-order representation theory of consciousness.

some sort. Higher-order theories hold that a mental state is conscious only if one is subjectively aware of oneself as being in that mental state – i.e., only in virtue of a "higher-order" state that is about the conscious mental state. In yet another formulation, according to higher-order theories, what makes a mental state a conscious mental state is the fact that it is accompanied by a simultaneous and non-inferential higher-order mental state whose content is that one is now in that mental state. The higher-order mental state can be a higher-order thought (HOT) or a higher-order perception (HOP), depending on the account[93]. According to the HOT version, for example, subjective awareness is due to one's having a thought that one is in the state. In other words, a mental state is conscious if one is subjectively aware of being in it (Rosenthal 2011, pp. 431 ff.).

Rosenthal developed the (HOT) in an attempt to operate a division of labour and investigate consciousness independently of intentionality and sensory character on the basis of the conviction that *we can* separate the question of what it is for mental states to be conscious from the question of what it is for mental states to have intentional or sensory character (Rosenthal 1990, p. 728). I agree with Rosenthal that we may separate the question of what it is for mental states (or conscious mental states) to be conscious from the question of intentionality, but I firmly believe – contrary to his view – that we cannot separate the question of what it is for mental states to be *conscious* from the question of the sensory character of conscious mental states. I believe that when it comes to consciousness a full division of labour is not possible and higher-order theories leave something behind.

Even if there have been attempts to defend the possibility of unconscious qualia (Shoemaker 1975, 1990; Nelkin 1989; Rosenthal 1991, 1997), I think that the very notion of unconscious qualia is incoherent (Papineau 2002). I fully agree with Ned Block (2011) that the HOT theory harbours an incoherence, namely that one could have a HOT without actually being in the qualitative state the HOT represents one as being in (see also Neander 1998, p. 420; Levine 2001, p. 108; Wilberg 2010)[94]. If one is a realist about the existence of conscious experience – as I am – the claim that one could have a HOT without actually being in the qualitative state the HOT represents one as being in is not an acceptable possibility[95].

---

93 For HOT see Rosenthal (1986, 1993, 2005); Gennaro (1995, 2004); Van Gulick (2000, 2004). For HOP see Armstrong (1968, 1984); Lycan (1987, 1996, 2004). See also Byrne (1997, 2001a, 2001b); Carruthers (1996, 2000, 2005); Dennett (1978a, 1991). For a general introduction see Van Gulick (2014, section 9.1); Carruthers (2011).
94 For a reply see Rosenthal (2011).
95 For extended criticism of higher-order theories of consciousness see also Dretske (1993; 1995).

Interestingly, this short excursus highlights that it is far from trivial that we all have a common background of assumptions regarding what "consciousness" means or what counts as "conscious". What I take to be the problem with higher-order theories of consciousness is that they seem to ignore the ontologically subjective dimension of conscious experience, which – in my opinion – is precisely what characterises consciousness and what makes the Natural Problem an interesting problem. In order to avoid any misconception regarding the notion of "consciousness" that I consider in the Natural Problem, in the next section I proceed to make explicit how I use the term "consciousness", the chief *explananda* of this research project.

4.2 Attempts to Characterize Consciousness

We have gone quite a long way without properly defining consciousness, but it is now time to do so. By proposing and endorsing a clear characterization of consciousness I want to ensure that everyone has the same object of research in mind and thereby preclude the possibility of misguiding and bewildering the reader with a vague use of the terminology. Hopefully this will remove a number of potential strawman objections based on misinterpretations of my claims and facilitate an accurate understanding of the philosophical discussion that is to come.

4.2.1 Consciousness: A Mongrel Concept

"Consciousness" and its cognates are widely used terms that occupy a vast conceptual space, connoting several different concepts and denoting different phenomena often partially or totally conflated with each other both in every day language and in the academic literature. Ned Block (1995, p. 277) famously highlighted this problem by calling consciousness a "mongrel concept" and claiming that there are a number of very different "consciousnesses" that ought to be sorted out.

One way to start off is to give a general characterization of the notion of consciousness as Searle (1990, p. 635) does:

> By consciousness I simply mean those subjective states of awareness or sentience that begin when one wakes in the morning and continue throughout the period that one is awake until one falls into a dreamless sleep, into a coma, or dies or is otherwise, as they say, unconscious.

This is quite vague, but it does give us a rough idea of what is usually meant by "consciousness". A similar characterization can also be found in the work of psychologist George Trumbull Ladd (1909, p. 30):

> What we are when we are awake, as contrasted with what we are when we sink into a profound and perfectly dreamless sleep or receive an overpowering blow upon the head – *that* it is to be conscious. What we are less and less, as we sink gradually down into dreamless, or as we swoon slowly away: and what we are more and more, as the noise of the crowd outside tardily arouses us from our after-dinner nap, or as we come out of the midnight darkness of the typhoid fever crises – *that* it is to become conscious.

The above are not proper definitions. They are simple characterizations entirely based on examples of what it is to be, or to become, "being conscious". The aptness of these characterizations to describe consciousness depends crucially on the subjective experience of the reader. If the reader knows what the (conscious) contrast between being awake and in a dreamless sleep consists of, then the characterization would help him get a grasp on what consciousness is. On the contrary, if the reader does not know what the (conscious) contrast between being awake and being in a dreamless sleep is (because, say, it is a non-conscious being), then the characterization alone would be insufficient to help that reader understand what consciousness is. This sort of rough characterizations is nonetheless useful since it ensures that the notion of consciousness is not confused with similar ones such as "conscience", "self-consciousness", or "cognition" (cf. Searle 1992, p. 83).

Searle claims, and I agree, that we cannot give a noncircular verbal definition of consciousness. In other terms, the only way to define a phenomenal (experience-based) concept such as consciousness is by making reference to other similar concepts and *vice versa*, and that is where the circularity arises. We can only define "consciousness" as "subjective awareness", "feeling", or something of the sort because any attempt to define consciousness in terms of non experience-based concepts (e.g., solely in terms of behaviour or objective facts) would fail to account for the experiential or subjective aspect that is essential and that defines the concept of consciousness[96]. This does not come without problems. It is not clear what notions such as "subjective awareness" or "feeling" mean and, as for consciousness, we cannot define them solely in terms of non experience-based concepts. Thus we are stuck in a definition loop where "consciousness" can only be defined by referring to other phenomenal concepts that, in turn, can only be defined by referring back to consciousness. The key to under-

---

96 This is the fault of higher-order theories highlighted in section 4.1.5.

standing all the above concepts and making sense of the characterization of "consciousness" as "subjective awareness" or "feeling" (and vice versa) is to *be* conscious. The circularity of the definition is not a deep problem for a being that can use it as a practical indicator of the phenomenon itself – consciousness *for* the being. Circularity is a serious problem only for someone who is not conscious and therefore has no grasp whatsoever of the core meaning of "consciousness", "subjective awareness", "feelings", or any other phenomenal concept. If one does not have a first-hand grasp on what phenomenality is in the first place, there is no way to get rid of the *impasse*. Any definition or characterization of phenomenal concepts (e.g., consciousness) in terms of non-phenomenal concepts would in fact be inadequate and incomplete. Moreover, any definition or characterization of phenomenal concepts in terms of other phenomenal concepts (e.g. "consciousness" as "subjective awareness") will be unintelligible because these individual terms refer to phenomenal concepts that the non-conscious being cannot understand precisely because of their phenomenality.

I think that general characterizations such as the ones above are useful to introduce the concept of consciousness. In consciousness literature, however, this has not always been deemed necessary. Some think that consciousness is all too familiar to be given any sort of definition. Psychologist George Stout, for example, held that properly speaking a definition of consciousness is impossible, but that that is not necessary since "everybody knows what consciousness is because everybody is conscious" (1899, p. 7). Similarly, William James never attempted to define consciousness in his *Principles of Psychology* (1890) by virtue of the fact that it is a too familiar phenomenon that is accessible by introspection and therefore needs no definition. A similar approach can be found in Freud, saying "what is meant by consciousness we need not discuss; it is beyond all doubt" (1933, p. 70). More recently, neuroscientists Francis Crick and Christof Koch (1990, p. 263), in an attempt to lay out the foundations of a neurobiological theory of consciousness, hold that a precise definition of consciousness is not needed since "everyone has a rough idea of what is meant by consciousness". I think that we should not take for granted that just because everyone is (or appears to be) conscious, everyone means the same thing when talking of "consciousness". More than this, I believe that any work on consciousness underestimating the need for at least a rough characterization of the explananda is responsible for the confusion that still reigns today in consciousness studies as to what the term "consciousness" denotes.

Some people however hold that we cannot give a clear and generally accepted definition of consciousness because there is no stable pretheoretical conception of it (cf. Titchener 1915, pp. 323–324). According to this interpretation, as Güzeldere (1997, p. 8.) puts it, consciousness "is a source of obscurity that remains typically recalcitrant to systematic investigation". A famously pessimistic and rather amusing definition of consciousness that goes in this direction is the dictionary entry "consciousness" by Stuart Sutherland:

> The having of perceptions, thoughts, and feelings; awareness. The term is impossible to define except in terms that are unintelligible without a grasp of what consciousness means. [...] Consciousness is a fascinating but elusive phenomenon: it is impossible to specify what it is, what it does, or why it evolved. Nothing worth reading has been written about it. (Sutherland 1995, p. 95)

I believe that this pessimistic attitude towards both the possibility of defining and studying consciousness is unjustified. Even though it is not possible to give a non circular definition of consciousness that is intelligible to one that does not know what consciousness is by means of experience, it is possible to characterize it in a way that is sufficiently precise to point always at one and the same phenomenon. I think we should be reasonable when dealing with consciousness: since there are good reasons to suppose that the philosophers and scientists that are interested in better understanding consciousness are themselves conscious, and that their consciousness does not differ radically, then there is no further reason to presuppose that there is no way to talk about consciousness in a way that makes sense. Characterizing consciousness in a way that is clear enough to define what we mean by it (and, importantly, what we do *not* mean by it) is valuable and required because that is the starting point for effective communication (i.e., avoiding misunderstandings) across and within consciousness-related disciplines. As Güzeldere says (1997, p. 8):

> If we hope that anything toward a better understanding of consciousness will come out of the joint efforts of different disciplines, it is of utmost importance to minimize crosstalk and make sure that common terms actually point to the same referents.

4.2.2 Toward a Taxonomy of Intransitive Being Consciousness

I take it that the above characterizations by Searle and Trumbull Ladd are good indicators of what "being consciousness" refers to, roughly speaking. Nevertheless, I want to further refine these characterizations by formulating a simple and unequivocal definition of "consciousness" that can easily highlight the crucial distinction between beings that are conscious *simpliciter* and beings that are unconscious *simpliciter*, regardless of further details about the intentional content or phenomenal quality of their global conscious state of mind. I want to formulate a definition of consciousness such that, with such a definition to hand, for any being S we could ask "is S conscious *simpliciter*?" and, if rightfully informed, it would be possible to reply with only two mutually exclusive answers: either yes (S is conscious *simpliciter*) or no (S is not conscious *simpliciter*).

With such a clear definition to hand it would be in principle possible to construct a *sensible coarse-grained conditional* taxonomy of intransitively conscious and unconscious beings. I suggest that this taxonomy would be "sensible" and "coarse-grained" because it would only aim at distinguishing beings that are conscious or unconscious *simpliciter*, rather than ambitiously aiming at classifying beings into fine-grained categories indicating quantitative or qualitative variations in the degree of consciousness (e.g. "S1 is a little more conscious than S2, and a little less than S3"), or indicating the targets *of* which different beings are or can be conscious. A coarse-grained taxonomy is a valuable goal because by virtue of being only modestly pretentious it has a relatively better chance of being correct in mapping the territory and furthermore it avoids giving us the unjustified impression that we know more than we actually do – something that socratically speaking we should always avoid. I suggest that the taxonomy would be "conditional" in the sense that its goal would not be that of classifying beings in clear-cut categories once and for all, but rather to say something like this: "if at time t – given our best theories – there are good reasons to think that being S1 has property C and being S2 has not property C, where C defines consciousness, then at time t we can conditionally classify S1 as conscious and S2 as unconscious". The only realistic goal is in fact to construct a taxonomy accounting for the most likely (i.e., most probable) state of affairs given the present knowledge – definitive knowledge being a chimera. With a clear definition of consciousness at hand, depending on whether our best theories suggest that a being meets the criteria stipulated by the definition of consciousness, we could conditionally classify beings either under the heading "conscious beings" or under the heading "unconscious beings". This taxonomy would be a faithful, informed, and up-to-date indicator of the state of affairs and it would be sensitive to the variation of our best scientific hypothesis about the distinction between conscious and unconscious beings (and, in the absence of evidence, sensitive to the most widespread folk intuitions about the same distinction). On the basis of new data it would be possible to re-evaluate whether a being meets the definition criteria and, if something changes, to review the taxonomy accordingly. The problem arising, however, is that it is not immediately clear from where we should start in order to formulate the definition of consciousness on the basis of which we could then delineate such a taxonomy of conscious beings.

4.2.3 Candidate Feeling Beings: A Problem

It might be tempting to think that in order to get a good definition we have to find our way by *first* listing the beings that we consider as exemplifying consciousness (e.g., conscious beings such as, supposedly, you), looking for a property that they have in common, and then control whether the beings that we do not consider as exemplifying consciousness (e.g., non-conscious beings such as rocks) lack such property. Consciousness could be tentatively defined as the differing property between what we take to be conscious beings and unconscious beings. This would be somewhat like defining what "red" by exemplification, that is, by saying something like "red is a property that strawberries, tomato, cherries, blood and Ferrari cars have in common, and that oranges, sunflowers, grass, the sky, violets, snow or coal do not have". With this strategy, the more complete the list, the greater the chance of reaching a fairly accurate characterization of what "red" is. However, adopting this strategy to define consciousness is problematic. The problem lies in the fact that – unlike "redness" – there is a more or less widespread disagreement when it comes to classifying beings (at least, some of them) into clear-cut mutually exclusive categories such as "conscious" and "unconscious". Regardless of one's *definition* of consciousness, there are clashing intuitions regarding which being legitimately counts as conscious, and which does not. Nobody would disagree that there are several important overall differences between – say –, you, a rock, a computer, a plant, an insect and a bonobo. However, it is not a unanimously apparent and acknowledged epistemically objective fact whether consciousness is amongst the properties of a rock, a computer, a plant, an insect and a bonobo.

What we usually do when attributing consciousness to beings other than ourselves is proceeding by means of unverifiable comparative deductions, such as deducing a similitude in consciousness between beings S1 and S2 from a behavioural, functional or anatomic similitude between S1 and S2. Since there is no way to verify whether consciousness really is amongst the properties of other beings (i.e., whether it is true that other beings that we take to be conscious actually are conscious), we cannot adopt this approach to construct a definition of consciousness. If I used plants as an example of conscious beings, for example, I would probably raise some confusion: some people might agree, some would assume that I am using "consciousness" incorrectly to mean something else, and some would review their notion of "consciousness" so as to make it fit what they think about plants. This does not mean that there is no common agreement at all. Indeed, most people would probably agree that humans are conscious beings whereas rocks are not. However the point here is that we are not justified in starting off with mere assumptions about which beings we think are conscious and which we think are not conscious in order to deduce a satisfactory characterization of "consciousness".

I think that the best way forward consists in dropping this "infra-beings" comparative method and acknowledging that the only sound – or at any rate, least controversial – way of suggesting a fair characterization of consciousness consists in drawing exclusively from our own subjective experience. In fact, given that we are conscious beings, and given that we can be both in conscious and unconscious states of mind, we are in a privileged position to suggest a reasonable definition of what distinguishes cases in which we are conscious from cases in which we are not conscious. The characterization of consciousness that I am aiming at, in short, has to be the *minimal sufficient condition* that is always verified when you would characterize yourself as conscious and that is never verified when you would not characterize yourself as conscious[97]. A good characterization of consciousness has to be formulated in such a way that the proposition "S is conscious" is universally true every time that, and only if, S is actually conscious.

4.3 Defining Criteria for Being Consciousness

There are three candidate characterizations of "being consciousness" that I think are worth considering: *consciousness as sentience and responsiveness* (section 4.3.2), *consciousness as wakefulness and normal alertness* (section 4.3.3), and *consciousness as what it is like to be a being, for the being* (section 4.3.4). On the basis of a warning on the limits of behavioural and neural characterizations of consciousness (section 4.3.1) I introduce and evaluate them, concluding that the best-suited characterization for the purpose of expressing the ontologically subjective dimension of intransitive being consciousness is the last one. These three characterizations of consciousness are conceptually distinct and therefore ought not be implicitly conflated under the same heading "consciousness".

4.3.1 Limits of Characterizations of Consciousness

There are various concrete situations in which it is crucial to evaluate the degree of consciousness of a given individual – think for example of a doctor checking the victim of a car crash, or of an anaesthesiologist keeping track of a person

---

[97] Notice that this does not imply that explicit awareness of one's consciousness is required to be conscious.

undergoing surgery. Since there is no way to have direct subjective access to another person's mental states, in order to evaluate the degree of consciousness of others we have to rely exclusively on the presence of indirect evidence or signs of consciousness. Identifying such indirect evidence appropriately is crucial to evaluate accurately the degree of consciousness of other people, and – if necessary – to treat them accordingly.

The medical community has developed a number of practical clinical scales to assess the degree of post-traumatic unconsciousness (in humans) relying mostly on indirect evidence coming from behavioural responses to stimuli. An example of this is the Glasgow Coma Scale (GCS), a generally applicable scheme of assessment of the depth and duration of impaired consciousness and coma (cf. Teasdale and Jennett 1974). Interestingly, this scale is not conceived as an "all-or-nothing" scale allowing determination of whether one is conscious or unconscious *simpliciter*, but rather a scale that allows for variations in degree of intransitive consciousness:

> To be generally accepted, a system must be practical to use in a wide range of hospitals and by staff without special training. But the search for simplicity must not be the excuse for seeking absolute distinctions where none exist: for that reason no attempt is made to define either consciousness or coma in absolute terms. Indeed, it is conceptually unsound to expect a clear watershed in the continuum between these states. What is required instead is an effective method of describing the various states of impaired consciousness encountered in clinical practice. Moreover, this should not depend on only one type of response because this may, for various reasons, be untestable. The three different aspects of behavioural response which we chose to examine were motor response, verbal response, and eye opening, each being evaluated independently of the other. [...] This depends on identifying responses which can be clearly defined, and each of which can be accurately graded according to a rank order that indicates the degree of dysfunction. (Teasdale and Jennett 1974, p. 82)

The Glasgow Coma scale is conceived as a tool allowing one to easily and reliably assess and keep track of changes in the degree of conscious impairment of a patient in repeated bedside assessment by independently measuring three kind of behavioural response to external stimuli: motor responsiveness, verbal performance, and eye opening. The best response in each component of the Glasgow Coma Scale is recorded by means of an indicative score (see Table 2, adapted from Laureys et al. 2005; see also Laureys et al. 2002).

A patient scoring high (say, obeying commands, orientated, and showing spontaneous eyes opening) will be assessed as more conscious than a patient scoring low (say, localizing pain, confused and opening eye only to pain)[98]. It is

---

[98] Notice that whenever one or more behavioural responses are impossible to assess due to

easy to understand why having such a general epistemically objective scheme of assessment of the degree of consciousness is of interest for those that have to deal with patients whose consciousness is impaired: this scale allows a wide range of observers (typically, doctors and nurses) to easily and impartially record behavioural responses with a high degree of consistency, thereby bypassing informal personal interpretations (i.e., epistemically subjective opinions on whether one is conscious) and avoiding biased and unjustified descriptions of the patient's global state of mind. Only epistemically objective behavioural signs are recorded – not the conclusions one may draw from them in terms of consciousness attributions.

Table 2

Motor Response (M)	6. Obeys commands 5. Localizing pain 4. Withdrawal from pain 3. Stereotyped flexion to pain 2. Stereotyped extension to pain 1. No motor response
Verbal Response (V)	5. Orientated 4. Confused 3. Inappropriate words 2. Incomprehensible sounds 1. No verbal response
Eye Opening (E)	4. Eyes open spontaneously 3. Eye opening to verbal command 2. Eye opening to pain 1. No eye opening

This is all reasonable and welcome, but it must not be forgotten that even if behavioural evidence is a handy clue to indirectly keep track and ultimately *infer* something about a person's consciousness from an external perspective, it is not properly speaking *proof* of one's consciousness, and certainly it is not consciousness itself. Similarly, it is worth highlighting that even though, for pragmatic reasons, we may deductively ascribe consciousness in general or particular

---

intubation, ventilation, or sedation this is registered as "not assessable" and therefore does not contribute to the total score. For medical purposes, the total score (up to 15/15) is less important than the specific combination of the individual responses.

conscious mental states to people on the basis of information about their neural states based upon a reliable account of NCCs, NCCs are not to be taken as a *proof* of the person's consciousness or as consciousness itself. NCCs are just ontologically objective data on the basis of which we can indirectly *infer* the existence of an ontologically subjective phenomenon. Whether or not this is actually the case, we cannot know for sure.

I am not saying this to undermine the empirical and methodological value of the above-mentioned scientific and medical practices (which, I repeat, are practically effective). I am saying this to remind ourselves that a satisfactory characterization of consciousness has to be a characterization of the ontologically subjective phenomenon itself, and cannot be restricted solely to a characterization of the ontologically objective ways in which the phenomenon is manifested externally and by means of which in practice we deduce whether one is or is not conscious. Basing our definition of consciousness only on the above mentioned criteria would miss the subjective nature of consciousness, and would therefore be an unsatisfactory definition.

I think that the attempts to characterize consciousness as "sentience and responsiveness" (section 4.3.2) or as "wakefulness and normal alertness" (section 4.3.3) suffer precisely from this problem. In these cases the way in which we individuate consciousness in others, with behaviourally identifiable and objectively quantifiable criteria such as sentience, responsiveness, wakefulness and alertness, overshadows the ontologically subjective dimension of consciousness, leaving no sufficient room for it in the characterization of the phenomenon. These characterizations do not stem from subjective experience ("what consciousness is for me"), but rather are influenced and informed by the way in which we proceed to evaluate whether beings other than ourselves are conscious. This, I think, is wrong. The task we are presently concerned with consists in characterizing as precisely as possible what consciousness is *for* ourselves, and in this sense the ontologically subjective mode of existence of conscious experience cannot be neglected. The task of defining how consciousness is exhibited in others is a different one, and not one I am presently concerned with.

4.3.2 Consciousness as Sentience and Responsiveness

According to one suggestion, what distinguishes conscious and unconscious beings is sentience. As Armstrong (1978, p. 722) says, if there is mental activity – "a single faint sensation" – actually occurring in the mind, there is consciousness. Thus, a being S could be described as "being conscious" if sentient, i.e., if it is capable of sensing (and responding to) the world.

At first sight this might appear to be reasonable: when I am conscious I am capable of sensing and responding to the world and, conversely, when I am not conscious I do not seem to be sensing and responding to the world. Moreover, beings that arguably cannot sense the world, such as rocks, are typically not regarded as conscious, whereas we tend to attribute consciousness to beings that clearly sense and respond to the world, such as higher mammals. However, the story is not as simple as that.

It is reasonable to think that beings with dissimilar sensory organs and/or sensory capacities, such as humans, gorillas, bats, scorpions and robots or even individual beings of the same kind (e.g., two individual humans), although they all are capable of *somehow* sensing and responding to the world, do so differently. It is not clear whether *any* kind of sensing is sufficient for consciousness or whether only *some* kind of sensing is: indeed, it is conceivable that some beings might be capable of *somehow* sensing and responding to the world and yet we would not readily classify them as "conscious". There are no clear-cut natural boundaries suggesting which kind of sentience is sufficient for consciousness. In this context, establishing which amount and what kind of sensory access to the world is sufficient to make a being conscious becomes largely a matter of arbitrary deliberation. Consider for example a robot able to sense the world using laser detectors/sensors and to move around on the basis of such information. Is this kind of "sentience" sufficient to consider the robot a conscious being? Think of plants able to sense their environment and intelligently adapt to it. Are they conscious beings? Opinions on these matters are diverging, and there is no reason to expect that we will ever have a clear answer to these questions.

If in order to give a clear criterion we establish that all *somehow* sentient beings are conscious, no matter how remarkably loose the notion of sentience is, we would end up having a vast pool of conscious beings in our taxonomy (including also the artificial devices and plants described above). This may indeed be right, but as a hypothesis it is suspiciously *ad hoc* since we do not have any independent plausible reason to think that all those beings are indeed conscious (e.g., behavioural evidence does not support this hypothesis). Moreover, we would still need extra work on top of the sentience criterion to explain why despite the diversity existing among different beings in terms of sentience such beings should all be equally considered conscious. Do we really want to say that humans, gorillas, bats, scorpions and robots are all conscious *simpliciter* regardless of their differences in sensing just because they all have *some kind of* sentience? I believe this would be unjustified.

Since (i) what counts as sentience is not clear, (ii) sentience broadly defined is a feature shared by many beings, and (iii) a good theory has to account for the fact that – for example – human-like sentience is probably different from the one of bats, scorpions or robots, I think that claiming that mere (indistinct) sentience is sufficient for consciousness is a hazardous and unjustified move. Since the definition of consciousness considered exclusively as capacity to sense does not account for the diversity of modalities by which different beings sense the world, I think this definition is not satisfactory.

Notice that my worry is not based on the idea that humans are sentient in a special way, and therefore ought to be considered conscious at a higher degree than other kinds of beings. What I am claiming is simply that since it is conceivable that some, but not all sentient beings are conscious, we are not justified in deducing from the fact that we are somehow sentient *and* conscious that any other being that we consider as being somehow sentient is also conscious – this inference being fallacious. The definition of consciousness as sentience is not satisfactory because it implies, without any reasons for it, that all somehow sentient beings are conscious.

Furthermore, even if it were true that every sentient being is conscious (something that we do not know), knowing this would not help us *explaining* what consciousness is. In fact, admitting that we had a firm grip on the notion of sentience (which is not presently the case), if we were able to identify beings as sentient we could say that they are conscious, but we would still not be *defining* what it means for those beings to be conscious or, more in general, what consciousness consists of. The problem – introduced in the previous subsection – is that sentience and responsiveness are ontologically objective characteristics that do not convey the ontologically subjective dimension characterizing consciousness. Since even knowledge of ontologically objective facts supposedly correlating with consciousness (i.e., sentience and responsiveness) does not entail knowledge of the ontologically subjective facts themselves (i.e., what consciousness is for those beings), there is no reason to define consciousness in terms of sentience and responsiveness.

4.3.3 Consciousness as Wakefulness and Normal Alertness

An alternative attempt to define consciousness consists in claiming that what distinguishes conscious beings from unconscious beings is the fact of being "awake and normally alert". Rosenthal (1993/2005, p. 46), for example, says: "a creature's being conscious means that it is awake and mentally responsive". Following this intuition, a being S is conscious if awake and normally alert, whereas it is not conscious if asleep and/or not normally alert. Along these lines, for example, we could say that humans, gorillas and any other kind of being is usually conscious, except when sleeping, in a deep coma, or in cases of abnormal alertness.

Again, however, this comes not without problems. First, there are some fuzzy cases where a being that does not count as awake and normally alert might nonetheless count as *somehow* conscious. A dreamer for example is *not* awake (dreams occur paradigmatically during sleep) and not *normally* alert, and yet she can be somehow mentally awake and responsive, as when one incorporates the sound of the alarm clock into one's dream, or in the case of sleepwalking. According to the present definition we would have to say that a dreaming being is simply not conscious, but there is a sense in which a dreamer might be considered at least minimally or "unusually" conscious. Similarly, a hypnotized being or a being having hallucinations is not awake and *normally* alert and yet might be considered minimally or "unusually" conscious[99].

Furthermore, if taken literally, this wakefulness criterion for "being consciousness" would be difficult to apply to all those beings radically different from us (especially non-animals or animals lacking a sleep-wake cycle). Does this means that beings lacking a sleep-wake cycle simply cannot be conscious? Or are there other ways to count as "awake and normally alert", such as being somehow "switched on" rather than "switched off"? Imagine a sunflower following the sun during the day and resting during the night. Can we say that the sunflower is "awake and normally alert" just when following the sun? Although we could describe as "wakefulness" this reflex action intrinsically dependent on the presence of the sun above the horizon (an external factor), it clearly is a different sort of "wakefulness" from that we would attribute to humans and most other mammals. This account looks even more suspicious if applied to robots or computers. Imagine a computer that is switched on and does whatever job it was conceived to do, thus being somehow "awake and normally alert". Could such a being be counted as conscious? And, if so, could it be conscious in the same way as animals and humans? Solving this worry would require us to give a definition of "wakefulness" and "alertness" that is applicable to any sort of being.

If we were to define being "awake and normally alert" in looser terms in order to make it general enough to accommodate the above raised problems, there would be a different problem, similar to the one raised for sentience and responsiveness. It is conceivable that a being with wakefulness X is conscious, and a being with wakefulness Y is not conscious. Indeed, it seems reasonable

---

99 There are a number of studies describing patients as awake *and* unconscious, something that Merker (2007) labels "unconscious wakefulness". Damasio (1999, pp. 95–101) describes patients in the midst of an absent epileptic seizure as unconscious while awake. Similarly, Bayne (2007, p. 18) describes patients in a vegetative state as "awake", in the sense that their eyes are open as part of the sleep-wake cycle, but as not conscious.

to think that some beings that can be "awake" in some sense (think of flora and machines) are not conscious. The definition of consciousness as loose wakefulness is not satisfactory because it implies, without giving reasons for it, that any kind of wakefulness is sufficient to be conscious – this inference being fallacious. We cannot deduce from the fact that humans that are awake and normally alert *are* conscious that any *somehow* awake and normally alert being is conscious too. This account, as the previous one, does not account for the apparent presence of various kinds of wakefulness.

Notice that if being "awake and normally alert" is the condition for being conscious, this entails that consciousness can be switched on and off continuously, since a being can be awake, asleep, wake up again, and so on. This would imply that consciousness would not be something that persists throughout the lifetime of a being (not even in a minimal sense), but rather something that can be repeatedly acquired and lost. This is not an intractable problem, but it is interesting to notice that it would require a satisfactory definition of persistence of our nature or identity in time independent of the notion of consciousness.

How could we say that humans are conscious beings *simpliciter*, since as a matter of fact they spend a significant part of their lives sleeping and thus not being conscious? One way would consist in establishing that a being counts as a "conscious being" if it has the *disposition* to be awake and normally alert, so as to account for the fact that every being that has the disposition to be awake and normally alert is conscious, regardless of whether the disposition is actualised or not at the moment of a distinction. Although – say – I am awake and normally alert (thus conscious) in Switzerland at the same time that Frank Jackson is asleep (and thus non-conscious) in Australia, it would be silly to say that we are different kinds of beings, i.e., that I am a "conscious being" while he is an "unconscious being", and that – when the day/night situation is reversed – he becomes a conscious being and I an unconscious being. We could account for the intuition that Frank Jackson and I belong to the same kind "conscious beings" by saying that we both have the disposition to be awake and normally alert. We can then still distinguish specific cases in which this disposition is instantiated (i.e., cases in which a conscious being is actually conscious) and cases in which the disposition is not instantiated (i.e., cases in which a conscious being is not actually conscious).

A decisive problem for this account of "being consciousness" is that even if it were true that having the disposition to be awake and normally alert is a sufficient condition to be conscious in all possible scenarios (something that we do not know), knowing that beings have the disposition to be awake and normally alert would not suffice to *explain* what makes them conscious. In fact, assuming that we had a firm grip on the notion of wakefulness (something that is not the case), if we were able to identify beings as awake we could classify them as conscious, but that would not *explain* what it means for those beings to be conscious or, more in general, what consciousness is. The problem – again – is that wakefulness and alertness are ontologically objective characteristics that do not convey the ontologically subjective dimension characterizing consciousness. Since even knowledge of ontologically objective facts supposedly correlating with consciousness (i.e., wakefulness and alertness) does not entail knowledge of the ontologically subjective facts themselves (i.e., what consciousness is for those beings), there is no reason to define consciousness in terms of wakefulness and alertness.

4.3.4 Phenomenal Consciousness (What it is Like)

The third way of characterizing consciousness – the one that I favour – is based on Thomas Nagel's (1974) "what it is like" criterion. Following this criterion, a being S is conscious if and only if there is something *it is like* to be being S, something it is like *for* being S. In other words, a being S is conscious *simpliciter* if and only if it is in a phenomenally conscious state of mind (i.e., if it feels)[100]. Conversely, if there is nothing it is like for a being S to be, nothing it is like for the being, then being S is not conscious. That is, a being is not conscious *simpliciter* if it is not in a phenomenally conscious state of mind (i.e., if it does not feel).

As suggested in section 4.1, I reserve the term "feeling" exclusively to refer to the global phenomenally conscious experience that an individual being has at a given moment in time when in a globally conscious state of mind[101]. Every

---

[100] Notice that it is possible to adopt the "what it is like" characterization in the transitive terminology of a first-order account of "state consciousness" by saying that a being S is conscious *of* a sharp pain, redness, a piece of music or thirst only insofar as there is something it is like (e.g., "sharpainness", "redness", "musicness", or "thirstiness") for S to feel a sharp pain, to perceive redness, to hear a piece of music or to feel thirst – something it is like for the being.

[101] A being S is conscious *simpliciter* if and only if it feels, where "feeling" (intended as verb) means experiencing in a general (i.e., not just in the "tactile" sense). I take expressions such as "I feel in love", "I feel angry", "I feel hungry", as synonymous with saying that there is something it is like for me to be in love, to be angry or to be hungry. I favour the reflective verb "feeling" over the expression "there is something it is like" because it is more concise and because the former sets the focus on the being's experience *simpliciter* (S feels), whereas the latter also hints at the intentional object or content of the being's experience (there is *something it is like* for S). Since for the Natural Problem I consider consciousness intransitively I think that characterizing the having of phenomenal consciousness as feeling ("ressentir" in French, "provare" in Italian) is a good choice: simple, intuitive, and straightforward. I adopt the noun-version "a feeling (sing.)/

feeling a being S has at a given time has necessarily some subjectively experienced phenomenal quality. I call this *global quale*, i.e., the *what* it is like to be "being conscious" for the being[102]. I call it this way in order to distinguish the *global qualia* of conscious global states of mind of a being from the classical use of *qualia* applying to particular conscious mental states. In the "state consciousness" vocabulary adopted by first-order theorists it is agreed that particular conscious mental states (e.g., a conscious pain) have a peculiar phenomenal character, quality, or *quale* (e.g., the "painfulness") that is consciously perceived by the subject having the conscious mental state. The "painfulness" of a conscious mental state of pain differs in phenomenal character or quality from the "sweetness" of a conscious perceptual experience of sugar, from the "redness" of a conscious visual perception of red, and so on. Since I am not assuming a "state consciousness" approach, I do not consider the distinctions between *qualia* of particular mental states of given beings. However, I believe that any conscious global state of mind of a given being necessarily has a *global quale* of some kind. That is, there can be no feeling that does not have some – a – global phenomenal quality (*global quale*), because that would imply that there is nothing it is like for the being to be in a given state of mind, which would in turn imply that the being is not conscious *simpliciter*[103]. Feelings can differ in terms of their specific

---

many feelings (plur.)" rather than the alternative "a feel (sing.)/many feels (plur.)" because I conceive a "feeling" as an event rather than as an object. However notice that most uses of the noun "feel" are interchangeable with my use of "feeling".

102 Kriegel (2003) claims that phenomenal character of phenomenally conscious experiences might consist in the compresence of qualitative character and subjective character. The distinction between subjective character (i.e., the being "for me") of an experience and its qualitative character, determining "what" the experience is like, is not uncommon in the literature (cf. Levine 2001, pp. 6–7). Representationalists such as Dretske (1995) and Tye (1995; 2000) seem to take qualitative character to be more fundamental, whereas higher-order theorists such as Armstrong (1968), Lycan (1990), Rosenthal (1990; 1991), and Carruthers (2000) seem to take the subjective character to be more important. I find most attempts of decomposing phenomenally conscious experience into separate but somewhat interrelated structural elements to be more puzzling than explanatory. Since it is possible to talk about phenomenally conscious experience without having to take a specific stance in such a debate on the structure of phenomenology I remain agnostic about this topic.

103 I am opting here for a quite uncontroversial thesis: if one (intransitively) feels it is conscious *simpliciter*; if one does not feel, it is not conscious *simpliciter*. It might however be possible to defend an alternative somewhat bizarre and demanding transitive theory postulating that it is possible to feel "nothingness". According to this theory, strictly speaking there are no unconscious beings: what we usually call unconscious beings are actually beings that feel exclusively "nothingness", which is qualitatively – but not *fundamentally* – different from other phenomenal experiences. Obviously, this can only work if there could be something it is like to feel "nothing-

global phenomenal character or quality (*global qualia*) – that is, *how* they feel like *for* the being, but – I claim – a being can only experience one *global quale* at a time. I believe in fact that it is phenomenologically sound to claim that what philosophers usually refer to as particular individual *qualia* are unified in actual phenomenology. That is, at any given moment in time, a being S who is in a phenomenally conscious state of mind has a single unified phenomenal spectrum/field – a single unified *global quale* – the *what* it is like to be for that being in that moment, which is not the mere sum of particular isolated *qualia*. We cannot genuinely consciously experience several *qualia* heterogeneously at the same time (e.g., "painfulness" + "coldness" + "redness" + "loudness" all at once). Rather, we have a single homogeneous phenomenal experience: a unified *global quale*[104]. In what follows I assume the view according to which conscious experience is phenomenally unified at any one moment in time.

Since I am interested in *intransitive* being consciousness, the specific *global quale* (i.e., the phenomenal content) of feelings does not matter when describing whether a being is "being conscious" or not. It is sufficient that a being experiences *some* – any – feeling, regardless of its *global quale*, in order to qualify that being as conscious *simpliciter*. This is important because, arguably, the quality of conscious experiences may differ severely across species, across beings of the same kind, and even between different experiences of an individual being (even though this cannot be empirically tested, since the "what it is like" is a feature that can be apprehended only via one's own personal phenomenal experience). In order to classify a being as conscious or unconscious *simpliciter*, having to consider *whether* a being feels, but not the *quality* or global *quale* of the feeling is a huge advantage. Humans can be described as conscious *simpliciter* because there are reasons to claim that there is something it is like for humans to be (i.e., humans feel). Similarly, if there are reasons to suggest that bats feel – i.e., that there is *something* it is like for bats to be and experience the world –, regardless of

---

ness" for the being – and this is controversial. The idea, however, would be that of viewing the absence of conscious experience as a peculiar kind of conscious experience, somewhat similarly to defining silence as a special kind of sound, rather than as the absence of sound. This might an interesting way to support the idea that all beings are always conscious – something that might be appreciated by panpsychists – and that for beings such as ourselves the only thing that varies is the quality of the experience rather than the very existence of experience (e.g., feeling "nothingness" when we are in a coma, and feeling in a different way – or something different – when we are awake and normally awake).

104 In chapter 7 I claim that global *qualia* might be subjectively phenomenally compared *a posteriori*, highlighting qualitative similarities across subjectively different globally conscious states of mind.

what it is like to feel for a bat, bats can be described as conscious *simpliciter*. This approach allows us to avoiding the notorious problem of knowing *how* feeling is like for other beings.

As Nagel (1974) famously held, even though we can try to imagine what it would be like to experience the world through echo-locatory senses as bats do, we cannot actually know what it is like to be a bat *for* a bat. This is because even if both bats and humans are conscious *simpliciter*, we are *differently* conscious *simpliciter*. Bats are bat-like conscious, and humans are human-like conscious. Bats feel in a bat-like way, and humans feel in a human-like way, each from their species-specific point of view. Even if humans had an echo-locatory system, it would not follow that our echo-locatory experience of the world would be the same or comparable with that of the bats. In fact, an echo-locatory system working as part of the human sensory system might yield different experiences to the same system working as part of a bat sensory system. Taking another example, it seems reasonable to suppose that what we experience when we smell roast beef is *somehow* qualitatively similar to what a bloodhound experiences, in that we both smell it. However, at the same time, we can suppose that by virtue of its superior olfactory receptors, the bloodhound's experience of that (or any other) smell could reasonably be richer or more fine-grained than ours. In sum, although both humans and bloodhounds might *somehow* smell roast beef, it is possible that we smell it *differently*[105]. Even if we can reasonably suppose that consciously smelling roast beef for a bloodhound is qualitatively similar to consciously smelling roast beef for us (even if the experience is perhaps richer in detail and more intense), we cannot know *how* roast beef actually feels like *for* a bloodhound. This suggests that however similar we might be, since every species and every individual being is restricted to experiencing only from its own subjective point of view, we cannot put ourselves in another being's phenomenological shoes. One's feelings may or may not be similar to those of other species or beings, but in *which* sense they could be similar is difficult to say. Even if from our own perspective we can try to *imagine* what consciousness could be like from any other subjective perspective, *de facto* we can only subjectively feel for ourselves, from our subjective standpoint. Given that in order to classify a being as conscious *simpliciter* we do not have to prove *what* the being feels, but only to give reasons suggesting *that* it

---

[105] See (Allen 1881) for a review of the different senses of smell in vertebrates. His intuition that dogs can discriminate odours at very low concentrations and that they are much times more smell-sensitive than humans is now empirically confirmed by studies (Coren 2004) estimating the olfactory cortex of the bloodhound dogs to contain nearly 300 million olfactory receptors against the 5 million of humans.

(probably) feels, this way of characterizing consciousness avoids a very insidious issue.

This is not to say that proving *that* a being feels is totally unproblematic. Indeed, given that our knowledge of consciousness is bound to our individual perspective, I agree that there is no way to decisively clear the road from incertitude and disagreement about *whether* gorillas, robots, scorpions and rocks are conscious *simpliciter* (even if we leave the issue of how they are conscious aside). However there is certainly less controversy in saying *that* a being feels on the basis of the evidence available, than in saying *what* the being in question feels.

Summarizing, I endorse the intransitive characterization of conscious beings as beings for whom there is "something it is like to be" or, better, as "feeling" beings, because this is an experience-centric definition granting a central and crucial role to the ontologically subjective dimension of consciousness. By so doing – that is, by laying subjective experience at the core of the characterization of consciousness – this definition highlights the ontologically subjective nature of consciousness, ensuring that what distinguishes conscious and unconscious beings is not an ontologically objective feature. Moreover, this characterization is indeterminate enough to leave an open room for the qualitative specificities of consciousness in different beings. What matters in classifying a being as conscious is only *whether* it feels, and this is perfectly compatible with the possibility that humans are human-like conscious, gorillas gorilla-like conscious, bats are bat-like conscious, and so on. An interesting feature of this characterisation of consciousness is that it accounts for the possible uniqueness of human consciousness, but it does not do so *a priori* at the expenses of different kinds of beings; it does not lean towards a detrimental species-chauvinism whereby only humans are or can be conscious, or conscious in the "right way". Any being that *somehow* feels is conscious, and what distinguishes some conscious beings from other conscious beings is just the way in which they are conscious, depending on what it is like to be for them.

4.4 Feeling and Self-Consciousness

In section 4.3 I have suggested that the best-suited characterization for the purpose of expressing the ontologically subjective dimension of intransitive being consciousness is that of "feeling being", i.e., that a being is phenomenally conscious *simpliciter* if it feels – if there is something it is like to be in a given state of mind for the being. Arguably, however, a being could be conscious also on a higher or meta-level: one could be self-conscious. The Natural Problem is a problem about phenomenal consciousness, and not about self-consciousness, but given the importance of the subject, it is useful to briefly explain what I take to be the relation between phenomenal consciousness and self-consciousness.

Roughly, a being is self-conscious if it is "aware of being aware" (Carruthers 2000), that is, if it is cognitively aware of itself is some way or another. This resembles in some sense the "wakefulness" consciousness, where a self-conscious being, instead of being only sense-wise awake, is also cognitive-wise awake: it is conceptually aware of its own mental states (e.g., particular thoughts, beliefs, perceptions, ...) or of its global state of mind.

There are various interpretations of what counts as self-consciousness, and which beings are considered to be self-conscious vary accordingly. Following the hard line of thought, a being is self-conscious *only* if it is explicitly conceptually aware of itself (or its existence). The fact that I think, "I am writing a dissertation", suggests that I am explicitly aware of my existence (or actions), and thus that I am a self-conscious being. If this is the criterion to determine which beings are self-conscious, most non-linguistic beings such as animals and probably young children would not count as self-conscious. On the other hand, if we lower the bar and allow for more rudimentary or implicit forms of self-consciousness (such as a non-conceptual self-awareness), then it might be possible to include more beings into the ranks of at least basically self-conscious beings.

I think that feeling (i.e., having phenomenal consciousness) and self-consciousness are distinct notions that ought to be kept conceptually distinct. Consider as an example the case in which you (phenomenally) feel a soft breeze on your back, but you are not cognitively aware that you are feeling it (i.e., you are not thinking "oh! A breeze!"). If something like this is possible – as I think it is – then one can be phenomenally conscious, and yet fail to be self-conscious. I keep the distinction clear by always using the explicit term "self-consciousness" when referring to higher-level reflective states of mind.

I have claimed that what distinguishes unconscious beings (e.g., rocks) from conscious beings (e.g., humans) is feeling, i.e., being phenomenally conscious. I am ready to grant that the additional presence of (conceptual) self-consciousness might allow us to distinguish beings with a twofold consciousness (feeling *and* self-conscious, such as normal human adults) from only phenomenally conscious beings (such as – arguably – scorpions, and – more controversially – newborn babies)[106].

---

[106] The reason for this is that newborn babies do not seem to be conceptually self-conscious as adults are. I use the expression "adult-like self-consciousness" interchangeably with "conceptual self-consciousness" to suggest that the latter is a sort of self-consciousness that is typically (and maybe only) possessed by human adults. By calling conceptual self-consciousness "adult-

I propose the notion "twofold consciousness" when referring to cases in which a being has both phenomenal consciousness *and* self-consciousness to suggest that even if the notions of phenomenal consciousness and adult-like self-consciousness have to be kept conceptually distinct, these phenomena might be intimately connected. In particular, within the diachronic approach that I am adopting, it might tempting to consider the hypothesis that the development of adult-like self-consciousness in phylogeny and ontogeny is somewhat connected with and consequent to the development of low-level phenomenal consciousness. It seems in fact plausible that, phylogenetically speaking, only-phenomenally conscious beings originated before beings with a twofold consciousness. In this sense, if we were to tackle a Natural Problem of *Self*-Consciousness, i.e., the problem of explaining why there are presently *self*-conscious beings, it might be necessary to dig further in the non-self-conscious mind first, and see whether only-phenomenally conscious minds hide the seed explaining the diachronic emergence of complex self-conscious minds[107]. I do not explore this possibility here[108], but tackling the Natural Problem of Consciousness – our problem – can be a first step in this direction.

Imagine a pyramidal structure regrouping the candidate characterisations of consciousness that we have considered, including self-consciousness. Every layer is a necessary, but not sufficient condition for all the layers that are set upon it. In our pyramid (Figure 2), the bottom layer is the *sentience* criterion. In fact, a being cannot be awake, phenomenally conscious, or self-conscious, without

---

like", I lay the ground for a non-conceptual minimal sort of self-consciousness possessed also by not-yet-adults that might explain how the adult-like comes to be.

107 To do so, it may be useful to consider what a feeling of self-consciousness consists of. It is not clear to me whether a feeling of ownership ought to be described as (i) a particular (token) feeling that is detached from – but works in conjunction with – other feelings, or (ii) a type of feelings that can come in many different forms, and whose quality varies coextensively with the sort of experiences one has. Imagine a self-conscious being looking at a red wall and having a feeling of redness as being *his own* feeling of redness. If (i) is the case, I would describe this situation as that of a being having the ownership feeling (Of) in conjunction with a specific feeling of redness (f_1). On the other hand, if (ii) is the case, I would describe the being as having a specific feeling of ownership about redness (Of_1). If now the being turns and looks at a blue wall, following (i) it would have the ownership feeling (Of) in conjunction with a specific feeling of blueness (f_2), whereas following (ii) it would have a specific feeling of ownership about blueness (Of_2).

108 I have raised this question in my 2010 Master of Letters Dissertation in Philosophy ("Can Effort Bridge the Fracture of Consciousness?") at the University of St Andrews under the supervision of Prof. Michael Wheeler.

also having some sort of primitive sense-access to the world[109]. Up one layer there would be the *wakefulness* criterion. In fact, it is hard to see how a being could be phenomenally conscious or self-conscious unless it is also "awake" in some fundamental sense.[110] On top of that comes phenomenal consciousness, i.e., what I take as defining "being-consciousness". In fact I think that a being cannot be self-conscious without being phenomenally conscious, albeit it could be phenomenally conscious without being self-conscious (as it is arguably the case for newborn babies). Finally, the top spot of the pyramid is occupied by (conceptual) self-consciousness.

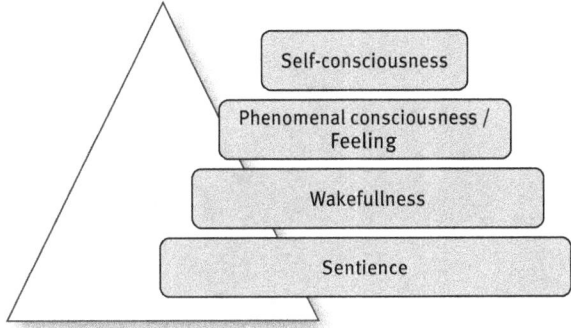

Figure 2

Considering the link between feeling and self-consciousness is an interesting project, but it extends far beyond the scope and the ambition of the present work. Even if self-consciousness is an important notion, it is not directly linked

---

109 Interestingly, most would agree that a piece of iron has no sense-access to the world and thus is never "wakefulness conscious". However, if you consider a robot exclusively made of iron that is able to interact with the environment in some rudimentary way – think of Brooks' can-picking robot Herbert (cf. Brooks 1991) –, when that "bunch of metal pieces" *does* something, it appears to us as a possible candidate to the title of "wakefulness conscious", when – really – it differs functionally, but not *intrinsically* from the piece of iron. Similarly, I take for granted that atoms have no sense-access to the world and thus are never "wakefulness conscious". However, humans, who are composed by atoms, undoubtedly *have* sense-access to the world. It follows that either we are mistaken in our presuppositions, or "wakefulness consciousness" would have to be some kind of emergent property.

110 Think of a dead man with no cerebral activity. Following the wakefulness criteria he is unconscious. I think it would be sensible to claim that such man is not self-conscious, nor can he have feelings, not even the feeling "what it is like" to be a dead human. Borderline cases such as sleep could be more problematic, since there is something that it is like to have a dream.

to the Natural Problem. I still marginally consider self-consciousness in the next chapter, but then I focus on feeling only since this is what the Natural Problem of consciousness is about.

In this chapter I have drawn some conceptual and terminological distinctions, suggesting what I mean by the expression "being conscious". On this basis of this work, in the next chapter I support the claim that consciousness presently exists and that it has been subject to a diachronic change.

5 Working Out Diachronic Claims

The main aim of this chapter is to back up two premises on the basis of which arises the Natural Problem of explaining *why* intransitive being consciousness presently exists, as I have argued in section 3.3. More precisely, the primary aim of this chapter is showing that there are good reasons to hold that (i) consciousness presently exists (i.e., that there presently are feeling beings), and that (ii) consciousness is subject to phylogenetic development. The secondary aim of this chapter is to show that – more in general – there are reasons to think that, in the actual world, phenomenal consciousness is subject to diachronic change both phylogenetically and ontogenetically. On the one hand, I do so to highlight the difference of claims arising if we clearly delineate the otherwise potentially ambiguous unspecified notions of "consciousness" and "development". On the other hand, I do so in order to highlight the vastness and interrelatedness of interesting consciousness-related questions arising within the diachronic framework I adopt.

I begin by defending the claim that feeling presently exists (section 5.1). Then, I distinguish two sets of diachronic claims that one could make about phenomenal consciousness: one about whether it is subject to phylogenetic evolution, and the other about whether it is subject to ontogenetic development. I distinguish "radical" and "qualitative" claims about diachronic changes of both sorts (section 5.2). Given this, I outline two different diachronic claims (section 5.3), and I individually consider them (section 5.4), concluding by summarizing the findings (section 5.5).

5.1 Feeling Presently Exists

I have claimed that feeling presently exists in the actual natural world, i.e., that it is a biological phenomenon that is presently instantiated. But is it really so? Does feeling really presently exist? In a naïve folk-psychological sense these questions might appear superfluous, but since they might potentially undermine an important claim on the basis of which the Natural Problem becomes salient, it is worth taking them seriously.

In chapter 4 I have defined feeling as the intransitive global phenomenally conscious state of mind of a given being at a moment in time. I have suggested that feelings can differ in their global *qualia* (i.e., *what* they globally feel like for the being in question), but that they all feel like something to the being. Accordingly, the necessary and sufficient condition for a being to count as a feeling being (i.e., an intransitively phenomenally conscious being) is to be able to have some (any) feeling. In other terms, a being S is a feeling being if and only if S can feel – regardless of what or how it feels. Conversely, a being that can have no feelings whatsoever is a non-feeling being (i.e., an intransitively unconscious being).

My claim is that there presently are feeling beings, and therefore that feeling – intended as biological phenomenon – presently exists. I think that we have good reasons to think that presently at least some beings (such as, supposedly, yourself) can feel, even granting that there presently are many non-feeling beings too (including non-feeling "objects", such as – supposedly – your shoes, standardly semantically opposed to feeling "subjects"). There are a few reasons that I take to be jointly sufficient to justify and defend such a realist position about the present existence of feeling beings and feeling intended as biological phenomenon. I explore them separately.

(i) *Phenomenological intuition:* Presently, any normal human being *does* feel. Regardless of what we might believe or conjure, this is a matter of fact. Pinch your arm, to use an example dear to John Searle, and you will feel pain. Watch a tomato and you will "feel" redness. Eat it, and you will feel "juiciness". Take hallucinogens, and you will feel "hallucinated". Feeling is a familiar and ordinary phenomenon. Since we feel and we are certain we do, there is a strong and compelling *phenomenological* reason to believe that feelings presently actually exist.

(ii) *The irreducibility of subjective ontology:* Feelings have a subjective ontology; that is, they exist only insofar as they are subjectively experienced. Regardless of whether one can do a psychophysical *causal* reduction of feelings to their neural supervenience base, it is not possible to reduce the subjective ontology of feelings to the objective ontology of concrete matter (see chapters 1 and 2). As Searle (2004, pp. 83) puts it, even if feelings can be entirely causally explained by neural behaviour, this does not come down to say that they are *nothing but* neuronal behaviour, since no reductive explanation in terms of objective facts could account for the additional fact that feelings have a first-person subjective character. Whatever feelings could be causally reduced to does not itself feel like anything. Given this and the phenomenological intuition outlined above, if we presently feel and the ontologically subjective nature of feeling cannot be questioned by means of an ontological (eliminative) reduction, then there is further support for the claim that feeling presently exist.

(iii) *Compatibility with solipsism:* One can hold a conservative epistemological view about other minds and might opt for epistemic solipsism, the view according to which one can only be certain of the existence of one's own mind, i.e., there is no way in which you can know for sure whether others also have minds (see section 3.2.1). Applied to feeling this position is – epistemologically speaking – a sound one. What it says is just that you are certain that you feel, but you cannot know for sure whether other beings feel too. This could lead one to hold that since we cannot know whether others feel, we ought to endorse the strong conservative ontological position claiming that beings other than yourself do not feel, i.e., ontological solipsism. Even if I have elsewhere argued that it is highly implausible that *really* I am the only feeling being and that everybody else is a non-feeling being, notice that ontological solipsism is in principle compatible with the claim that presently feeling exists, since for the claim to be true it is sufficient for one being to presently feel. If presently I am the only conscious being and you are a zombie, just trust me: I *really* do feel, and feeling presently exists. Similarly, if presently you do feel, regardless of whether anyone else actually feels, you have reasons to accept the claim that presently consciousness exists.

(iv) *Compatibility with fictional feelings:* Even if we were living in a Matrix-like fictional world, the fact that some or all our feelings are fictionally induced hallucinations or illusions would not change the fact that we presently *do* feel. Regardless of whether we presently have "fictional" feelings that are inappropriately connected (or misrepresent) ontological objective reality or "appropriate" feelings that are appropriately connected (or represent) ontological objective reality, we have reasons to think that we presently feel. We can doubt the reliability of *what* we transitively feel (i.e., whether what we are phenomenally conscious *of* is true), but we cannot doubt the very fact that we intransitively feel (i.e., whether it is true that we are phenomenally conscious *simpliciter*). Thus, even if our feelings – unbeknownst to us – presently misrepresent the world, having fictional feelings is nevertheless sufficient to claim that feelings presently exist[111]. I raise a similar point more in general in (v).

(v) *Argument against skeptics:* A skeptic can make a point by providing a scenario X, indistinguishable from reality Y, that can deceive us. For example, a skeptic about the external world could suggest that for every possible world W_1 where people experience a reality Y, there is a world W_2 totally subjectively indiscernible from W_1 where people live in a dream or a virtual reality X that they (mistakenly) experience as being the reality Y. The skeptic would then say that since W_1 and W_2 are subjectively indistinguishable we could be living in a type-W_2 world rather than in a type-W_1 world as we think we do[112]. Such

---

[111] I do not think that our feelings are fictional in this sense, but I grant that – at least as theoretical possibility – I could be wrong about that.

[112] This idea that has generated a quantity of science fiction works such as the films *The Matrix* (Andy and Lana Wachowski, 1999) and *Inception* (Christopher Nolan, 2010).

a situation, however implausible, is not logically inconsistent (after all there could be a big deceiver) and thus the skeptic argument is at least logically tenable. However, I argue, a similar position with respect to the very *existence* of feelings in the actual world is untenable. There are many metaphysically possible zombie worlds where there presently is no feeling (i.e., where there are no intransitively conscious beings), but our actual world *cannot* be one of them. It might be that what we describe as "redness" is really something else (e.g., "blueness"), and even that we are systematically or constantly mistaken about how the *quality* of our experiences represents the world, but this does not show that our feelings do not exist *at all*. In other terms, we might be mistaken about *what we think* we experience, but not about *the very fact that* we experience. The important point here is that regardless of whether our experiences are veridical or not, and regardless of whether our phenomenally conscious states of mind are natural or artificially induced, the existence of some (any) feeling is sufficient to counter the claim that feelings do not presently exist at all in the actual world.

A skeptic about the existence of feelings would have to find a gap for the deceiver to intervene between the external world and the internal first person (subjective) perspective. One could argue that we do not feel, and we only *think* with absolute certainty – but mistakenly – *that* we feel[113]. The very act of thinking something like that, however, feels like something, leading to a performative self-contradiction (see section 2.1.3). Alternatively, one could argue that a specific kind of feeling – say, the global quale of feeling the world by means of echolocation – does not exist in the actual world, i.e., that it is just a philosophical invention which is not instantiated in reality. After all, given that there is no way in which we can know (by introspection or other) whether bats or any other being can or cannot feel that way in the actual world, it might be reasonable to make such an ontological claim. However, even if such an ontological claim happened to be true, i.e., even if really there is no global quale of feeling the world by means of echolocation, this would not show that feeling *simpliciter* in general does not presently exist. There are plenty of feelings that we do experience in the actual world, and even just one of those suffice to hold the realist claim that feeling (in general) presently exists in the actual world. Anyone who denies the existence of feeling in the actual world has the burden of proof. But proving that we do not feel (or – arguably – that anything does not exist) is impossible. You might

---

[113] Higher-order theories of consciousness (section 4.1.5) are particularly exposed to this sort of worry.

manage to convince a non-conscious being (if it can be convinced of anything at all) that feelings do not exist, somewhat like a psychic could convince you that extrasensory perception does not exist – that is, supposing that you are not a psychic yourself. However, supposing that you happened to actually have extrasensory perception, you could not believe anymore the psychic telling you that extrasensory perception does not exist. Similarly, if you do feel, you cannot justify the claim that presently there are no feeling beings.

My ontological claim that feeling presently exist rests heavily on phenomenology-based remarks. This raises the question of whether and to what extent we are justified in relying on *our own* experience to assess ontological questions. It is conceivable that experience in adult human beings such as us differs from that of newborn babies, non-human animals, and other sorts of beings. Furthermore, it is conceivable that some people's epistemological intuitions about phenomenology might differ from someone else's intuitions. So, what reason is there to endorse the ontological conclusions resulting from *my* intuitions about feelings (and the intuitions of those who share them) rather than others? Granted that intuitions are not always trustworthy and reliable indicators of how the world is, my reply is that if we want to advance in the understanding of consciousness we have to be pragmatic and accept the limitation of having to introspectively rely on our experience, since this is the only *direct* source of information available (see also chapter 1). Either one bites the bullet and tries to understand something about consciousness despite these harsh epistemic conditions, or one surrenders to the limited epistemic tools we have at our disposal and joins the pool of skeptics and critics treating consciousness as a taboo topic. Since the only available *direct* evidence of the existence of feeling is self-centred, anthropocentric, phenomenological experience, I think it is reasonable to refer to that. I am confident that the modest ontological claim that there presently are intransitively conscious beings (i.e., beings that feel *simpliciter*) is uncontroversial enough to gather a vast consensus (among a presumably conscious audience) even if it relies on introspection. Of course, there might be more to consciousness that what we know about it from our adult human bound phenomenological; however – and this is the central point – there cannot be *less* to consciousness than what we know from such a perspective. What we know, i.e., *that* we presently feel, is sufficient to claim some existence, and – by extension – *the* present existence of feeling[114].

---

[114] It is of course problematic judging in a fine-grained manner which non-normal-adult-human beings feel. Due to our limited epistemic resources, we need some working criteria allowing us to establish in the least possibly controversial manner whether a given being fits or not the definitions of feeling being given above.

5.2 Phylogenetic Evolution and Ontogenetic Development

So far I have argued for the claim that (i) feeling exists. The second claim underlying the Natural Problem that I need to back up is the claim that (ii) feeling is subject to phylogenetic evolution. I begin by clearly outlining the differences between diachronic change claims about the phylogenetic evolution of phenomenal consciousness (section 5.2.1) and diachronic change claims about the ontogenetic development of phenomenal consciousness (section 5.2.2).

5.2.1 Phylogenetic Evolution

> *Phylogenetic Evolution General (PE general):*
>
> Phenomenal consciousness is subject to phylogenetic evolution, i.e., consciousness is a feature that some species developed during evolution.

Phylogenetics, literally meaning the origin/birth of races, is the study of the evolutionary relations among groups of organisms (e.g., species) in biology. By claiming that phenomenal consciousness is subject to phylogenetic evolution (i.e., PE general), I mean that it is a feature that at least some recent species in the tree of life – that is, recently evolved species – have, but that their ancestors either (i) lacked or (ii) had in a less developed way[115]. Notice that the claim (i), "ancestors of conscious species lacked consciousness," is different and much stronger than the claim (ii), "ancestors of conscious species had consciousness in a less developed way". What I mean by (i) is that the ancestors had no feelings at all. On the other hand, what I mean by (ii) is that although ancestors of conscious species had *some* sort of feelings, they did not have the same sorts of feelings that their descendants developed. I will come back to this more in detail. For the

---

[115] Remember that I take consciousness in the actual natural world to be a biological phenomenon, and this is why the species I consider are only biological species that live or have lived on planet Earth. This is not to be interpreted as some sort of bio-terrestrial chauvinism about consciousness, but rather as a direct consequence of the realist naturalist approach I am adopting in order to understand why consciousness presently exists in the actual world. My approach consists in focusing on those species that I take to have (or have had) consciousness in the actual world, i.e., some biological species that correspond to given criteria for consciousness, and to look at their phylogenetic history in order to explain why consciousness presently exists. My strategy consists in clarifying what the biological function of consciousness as we know it in the actual world might be. This might in principle then be used to tackle the additional question of knowing whether and how different sorts of beings (artificial, extra-terrestrial, and so on) are or could be conscious, but that is not my goal.

moment, I just want to highlight that the "PE general" claim includes two more specific claims, namely:

(i) *PE radical change*:

The fact that some beings are conscious is subject to phylogenetic evolution.

(ii) *PE qualitative change*:

The manner in which beings are conscious is subject to phylogenetic evolution.

5.2.2 Ontogenetic Development

Ontogenetic Development General (OD general):

Phenomenal consciousness is subject to ontogenetic development, i.e., phenomenal consciousness is a feature that some individual beings develop within their own lifetime.

Instead of a long-term species-related evolutionary perspective, one can adopt a short-term perspective covering only the development of phenomenal consciousness in an organism's lifespan. Ontogenesis, literally meaning the origin/birth of a being, is the study of the origin and the development (i.e., the developmental history) of an organism within its own lifetime[116]. By claiming that phenomenal consciousness is subject to ontogenetic development (i.e., OD general), I mean that phenomenal consciousness is a feature that some beings develop within their own lifetime. This claim can take different specific forms: either (iii) a being has no phenomenal consciousness at all when it is conceived and develops it during growth, or (iv) a being has some sort of phenomenal consciousness when it is conceived, but not the same sort of phenomenal consciousness that it comes to have at later stages in its life. Again, I will come back to this more in detail. For the moment, I just want to highlight that the general OD claim includes two more specific claims, namely:

(iii) *OD (radical change)*:

The fact that some beings are conscious is subject to ontogenetic development.

(iv) *OD (qualitative change)*:

The manner in which beings are conscious is subject to ontogenetic development.

---

[116] Again, the beings I consider are only biological individuals that I take to have consciousness, since this is the only viable way to study the ontogenetic development of consciousness.

There can be a link between ontogenetic development and phylogenetic evolution. Individual organisms undergo developmental processes – that is, ontogenetic processes – that can eventually trigger species undergoing evolutionary processes – that is, phylogenetic processes. What matters for the present case, however, is only that the two processes operate differently and are to be kept conceptually distinct. This is why I outlined the differences between claims about the phylogenetic evolution (PE) and claims about the ontogenetic development (OD) of phenomenal consciousness in their general and specific (i.e., "radical" and "qualitative") forms.

5.3 Two Fine-Grained Questions

The above distinctions suggest that phenomenal consciousness is subject to both phylogenetic evolution and ontogenetic development. Arguably, the same sort of claims could be made also in the case of *self*-consciousness (see chapter 4). I distinguish the resulting four fine-grained diachronic claims in Table 3, highlighting in grey the questions concerning self-consciousness that I do not tackle.

Table 3

	Phylogenetic Evolution	Ontogenetic Development
Feeling	(A) feeling is subject to phylogenetic evolution	(B) feeling is subject to ontogenetic development
Self-Consciousness	(C) self-consciousness is subject to phylogenetic evolution	(D) self-consciousness is subject to ontogenetic development

Opposed to these four claims, one can argue for their respective antitheses (Table 4).

Table 4

	Phylogenetic Evolution	Ontogenetic Development
Feeling	(¬ A) feeling is not subject to phylogenetic evolution	(¬ B) feeling is not subject to ontogenetic development
Self-Consciousness	(¬ C) self-consciousness is not subject to phylogenetic evolution	(¬ D) self-consciousness is not subject to ontogenetic development

Finally, one can remain agnostic as to whether any of the above is the case (Table 5). One can say that we simply (i) do not know, or – a stronger thesis – (ii) cannot know, whether feeling or self-consciousness evolved phylogenetically or developed ontogenetically.

Table 5

	Phylogenetic Evolution	Ontogenetic Development
Feeling	(A ∨ ¬ A) we do not/cannot know if feeling is subject to phylogenetic evolution	(B ∨ ¬ B) we do not/cannot know if feeling is subject to ontogenetic development
Self-Consciousness	(C ∨ ¬ C) we do not/cannot know if self-consciousness is subject to phylogenetic evolution	(D ∨ ¬ D) we do not/cannot know if self-consciousness is subject to ontogenetic development

To sum up, for each of the four above mentioned diachronic claims concerning phenomenal consciousness and self-consciousness there are three possible stances: a positive one, a negative one and an agnostic one (Table 6).

Table 6

Fine-grained questions	Positive answer	Negative answer	Agnostic answer
a) Is feeling subject to phylogenetic evolution?	A	¬ A	A ∨ ¬ A
b) Is feeling subject to ontogenetic development?	B	¬ B	B ∨ ¬ B
c) Is self-consciousness subject to phylogenetic evolution?	C	¬ C	C ∨ ¬ C
d) Is self-consciousness subject to ontogenetic development?	D	¬ D	D ∨ ¬ D

When formulating the Natural Problem (section 3.3) I have assumed that, in the actual natural world, the current existence of consciousness – intended as biological phenomenon – is the outcome of evolution by natural selection. In order to back up this claim, I need to give reasons in support of the claim that feeling is subject to phylogenetic evolution – the fine-grained claim (a). This is my primary goal. My secondary goal is to show that there are reasons to think that, in the actual world, phenomenal consciousness is subject to diachronic change both phylogenetically and ontogenetically. In order to achieve both goals, I also tackle (b), the other fine-grained question concerning feeling. I tackle these two claims separately explaining why I endorse the positive answers, i.e., answer (A) to question (a), and answer (B) to question (b). I argue that, among all the possible answers, these are the most promising and least controversial to endorse. I do not pursue further the issue of self-consciousness, but this might be an interesting topic for another work.

5.3.1 Radical Change Claim VS Qualitative Change Claim

In order to assess the correctness of (A), and (B) in a sound way, I need to account for the strength of the claim that feeling is subject to phylogenetic evolution/ontogenetic development. In fact, as I already mentioned, the strength of the claim can vary. On the one hand, I test whether species or beings are subject to a *radical change* regarding consciousness, i.e., if species or beings can go from not feeling at all to feeling. This is an "all or nothing" change in the *fact* of being conscious. On the other hand, I test whether species or beings are subject to *qualitative changes* regarding consciousness, i.e., if species or beings can go from having (only) rudimentary sorts of feelings to having different fully-fledged sorts of feelings. This is a change in the *manner* in which species or beings are conscious.

In order to render this fact/manner (or radical/qualitative) distinction more explicit I show how it applies to the phylogenetic evolution and ontogenetic development of phenomenal consciousness respectively.

Let S_1 be a species at t_1 and S_2 one of its successive generation at t_2:

(i) *PE (Radical Change)*:

The *radical change claim* about the phylogenetic evolution of phenomenal consciousness says that S_1 is unconscious, whereas S_2 is phenomenally conscious. *The fact that* species are phenomenally conscious is subject to phylogenetic evolution.

(ii) *PE (Qualitative Change)*:

The *qualitative change claim* about the phylogenetic evolution of phenomenal consciousness says that S_1 is *somehow* phenomenally conscious, but cannot have global qualia X, whereas S_2 can have global qualia X. The *manner* in which species are phenomenally conscious is subject to phylogenetic evolution.

Let S_1 be an individual being at t_1 and S_2 the same being at t_2:

(iii) OD (Radical Change):

The *radical change claim* about the ontogenetic development of phenomenal consciousness says that S_1 is unconscious, whereas S_2 is phenomenally conscious. *The fact that* a being is phenomenally conscious is subject to ontogenetic development.

(iv) OD (Qualitative Change)

The *qualitative change claim* about the ontogenetic development of phenomenal consciousness says that S_1 is *somehow* phenomenally conscious but cannot have global qualia X, whereas S_2 can have global qualia X. The *manner* in which a being is phenomenally conscious is subject to ontogenetic development.

For both phylogenetic evolution and ontogenetic development, both the radical change claim and the qualitative change claim are individually sufficient but not individually necessary to argue that phenomenal consciousness is subject to diachronic change. If the radical change claim succeeds – that is, if there has been a radical change in the *fact* of having phenomenal consciousness in the history of a given species/being – the qualitative change claim – saying that there has been a historical change in the *manner* in which that given species/being is phenomenally conscious – is unnecessary for the claim that consciousness is subject to development. Conversely, if the qualitative change claim succeeds, the radical change claim is unnecessary for the claim that consciousness is subject to development. However, we can only claim that phenomenal consciousness is subject to diachronic change if *at least* one of these claims is true. In other terms, it is necessary that either the radical change or the qualitative change be true to hold that phenomenal consciousness is subject to diachronic change.

To put it formally:

(Phenomenal consciousness is subject to diachronic change) if and only if [(radical change) ∨ (qualitative change)]

In order to examine whether phenomenal consciousness is subject to some kind of diachronic change (i.e., phylogenetic evolution or ontogenetic development), I first consider whether the *radical change* claims are correct. Then I consider whether the *qualitative changes* claims are correct. If both the radical and the qualitative change claims turn out to be incorrect or highly doubtful – i.e., only if phenomenal consciousness turns out to be neither the subject to radical change nor to qualitative change in time – this would suggest that phenomenal consciousness is not subject to diachronic change. In all the other cases, there are reasons to believe the claim that phenomenal consciousness is subject to diachronic change. In the next section I apply this strategy separately to the fine-grained claims (a) and (b) outlined above.

5.4 Examining the Two Fine-grained Diachronic Claims

5.4.1 Feeling Is Subject to Phylogenetic Evolution (a)

The *radical change claim* about the phylogenetic evolution of feeling acknowledges that there are species S_2 that presently feel (among which *homo sapiens*), and holds that they all have some unconscious ancestor species S_1. Is this the case?

The positive answer, (A), says that – phylogenetically speaking – there have not always been feeling species in evolutionary history. Indicatively, feeling species started to exist when the first neurophysiological structures suitable to sustain feelings evolved. If you believe that the very first species that lived on our planet could not feel and you agree that amongst their descendants there are now species that feel, you are buying into this thesis.

The negative answer, ($\neg$ A), says that, acknowledging that there are feeling species today, phylogenetically speaking there has not been any kind of evolution in the *fact* that species feel. In other terms, if we could track down any ancestor species of presently feeling species back to the beginning of evolution we would see that that was a feeling species too. According to this thesis feeling species have been around since the beginning of life on the planet.

The agnostic answer, (A $\vee$ $\neg$ A), says that we (i) do not know and/or (ii) cannot know whether the very first species that lived on our planet could or could not feel. As a consequence, we cannot argue for either (A) or ($\neg$ A) because doing so requires taking a stance on whether species could or could not feel – which we cannot know.

This last position is the most waterproof as far as epistemology is concerned. In fact, up till now, there is no proof suggesting whether or not our ancestors could feel. Moreover, since feeling is an ontologically subjective feature of the world, it seems reasonable to question whether (or how) such proof could be found at all. It is already very difficult to agree on whether most of the beings alive today feel, despite the fact that we can monitor them extensively and – in some cases – we have access to subjective reports. Assessing whether species that do not exist anymore could or could not feel seems therefore an almost impossible challenge[117]. This, however, is only an epistemological problem. As far as

---

[117] Feinberg and Mallatt (2016), relying on fossil record of evolution and on a list of biological and neurobiological features considered responsible for consciousness, suggest what might

ontology is concerned, either (A) or (¬ A) has to be true. In other terms, it may be true (and indeed it probably is the case) that we will never *know* for sure which one among (A) or (¬ A) was the case, but it is necessarily the case that for every species in the evolutionary history either it felt or it did not, since these are mutually exclusive options[118]. There are no knockdown arguments to hold either one of (A) or (¬ A) over the other. Nonetheless we can attempt to settle this ontological question on the basis of our definition of feeling.

A good starting point to argue in favour of (A) consists in looking at species alive today and seeing how they relate to the last universal common ancestor – the most recent common ancestor from whom all organisms living now on earth descend. Arguably, some species alive today can feel (e.g., humans, higher primates, and so on), whereas others cannot presently feel (e.g., bacteria). Now, either (i) the last universal common ancestor *could not* feel and some of its descendent species *evolved* that trait, or (ii) the last universal common ancestor *could* feel and some of its descendent species *lost* that trait. If (i) is correct, then we can conclude that feeling is subject to a "positive" radical change during phylogenetic evolution because specie S_1 was unconscious, whereas some present species S_2 can feel. The same is the case if the opposite (ii) is correct: if S_1 could feel, but some S_2 are presently unconscious, feeling is subject to a "negative" radical change in phylogenetic evolution.

Although not impossible, it is highly improbable that (ii) is the case. The reason for this is simply that the last universal common ancestor was a physiologically simple being such as – or even simpler than – the beings alive today that are usually considered as non-feeling beings (e.g., bacteria). This suggests that if we agree that bacteria and other such simple beings are unconscious in virtue of their insufficient neurophysiological architecture (i.e., insufficient as supervenience base for feelings), then – by the same standards – there is no reason to

---

have been the case. Among other things, they suggest that consciousness might have appeared about 520 to 560 million years ago, when the great "Cambrian explosion" of animal diversity produced the first complex brains. From this they suggest that all vertebrates (including fish, reptiles, amphibian, and bird) are and have always been conscious, and that arthropods (including insects and probably crustaceans) and cephalopods (including the octopus) meet many of the criteria for consciousness.

118 I think that the presence of feeling is clear-cut once we have defined (i) what we mean by feeling, and (ii) which criteria we use to determine whether a being fits that definition. This of course does not mean that it is epistemically simple or clear-cut to draw the distinction: it clearly is not. What I want to say is that such a distinction is in principle possible. In other terms, if we had all the information needed to evaluate the presence of feelings against a standard of what counts as feeling, we could do that.

suppose that the last universal common ancestor could have been phenomenally conscious. This, of course, is tied to the naturalistic conception I am endorsing suggesting that a being's ability to feel is closely constrained by its physiological structure and organisation. The reason why we can be almost sure that bacteria cannot feel is that they lack any sort of functional equivalent to a nervous system that could enable them to have such feelings in the first place. Similarly, there are overwhelmingly good reasons to hold that the last universal common ancestor – that was supposedly as simple as bacteria – lacked the minimal neurophysiological architecture required to feel.

Under my assumptions the species-specific neurophysiological equipment is *de facto* the result of phylogenetic evolution (e.g., humans have a central nervous system, whereas other species lack it). Differences in neurophysiological architecture are decisive in influencing the *fact* that some species can feel whereas others cannot. Therefore, at least indirectly – through the evolution of the physiological structures allowing some beings to feel – feeling seems to be subject to phylogenetic evolution. I believe this to be rather uncontroversial, and thus a good reason to claim that something along the lines of (i) must be the case, i.e., that feeling is (at least indirectly) subject to a radical change during phylogenetic evolution.

The only remaining objection suggesting that the radical change claim is mistaken would consist in claiming that neither (i) nor (ii) is correct. One could argue against our criteria for a third hypothesis saying that (iii) all the presently living species can feel, and so could the last universal common ancestor too. In other terms, one could suggest that all the living beings, both of the past and of the present, are phenomenally conscious *simpliciter*. If this is right, then feeling would not be subject to radical change in phylogenetic evolution. It seems – however – that we can only grant feelings to the last universal common ancestor by significantly altering the general standards for what counts as a feeling being. This would also entail that species such as bacteria would also have to be considered as feeling species. This move would clearly be *ad hoc*, and would encounter a number of problems. The main objection to this move would be to say that such an alteration of standards for the attribution of feeling has no reason to exist in the absence of evidence, since it would arbitrarily violate our folk-psychological and available scientific evidence suggesting what counts as a feeling being. I think that this objection is sufficient to suggest that (iii) is misleading and that the radical change suggesting that feeling is subject to radical phylogenetic development is correct[119].

---

119 See also my objection to panpsychism in section 3.2.2 for additional reasons.

However, since (iii) could cast a doubt upon the radical change claim, I show that that even if feeling were not subject to radical change (which seems to be rather unlikely), we could still make sense of the claim that feeling is subject to phylogenetic evolution. I do so by going over to the second test, the *qualitative changes claim* about the phylogenetic evolution of feeling. This test says that if a species S_1 can feel, but cannot have global qualia X whereas – instead – S_2 can have global qualia X, then, feeling is subject to *qualitative* phylogenetic development. There are reasons to think that even if all species could feel *simpliciter*, they would feel *differently*. Depending on their embodiment, neurophysiological architecture, environment and so on, different species would be bound to feel in different manners. It is absurd to think that the last universal common ancestor could have had any sort of global qualia despite its physiological simplicity and the lack of the perceptual apparatuses permitting, for example, visual or auditory experiences. It is even more absurd to hold this and claim that during speciation some of these feeling capacities got lost here and there. In other words, even admitting that the last universal common ancestor could *somehow* feel, clearly it could not have felt in the same manner as presently existing species S_2 can feel. If this is true, then feeling is at least subject to *qualitative* phylogenetic development. In other words, at least the *quality* of species-specific feelings is subject to diachronic change throughout phylogenetic evolution. I believe that this point in favour of (A) is strong enough to settle the question; even more so if supported by the radical change claim outlined above.

Let us nevertheless look at possible arguments in favour of (¬ A) suggesting that also the qualitative change claim is mistaken, i.e., suggesting that phylogenetically speaking there has not been any kind of change in the *manner* in which historically related species feel. Any such claim relies on the thesis that all species and their ancestors are equal with respect to *how* they feel, i.e., that feelings are a fixed feature of the world that is not subject to qualitative phylogenetic evolution. One extreme – and extremely unlikely – option would be to argue that no species have ever felt and that no species presently feel. If that were true, then it would be true to say that there is no qualitative difference in how different species feel because, in fact, all species do – equally – not feel anything at all. This is a position that nobody, I think, would seriously hold, since it is bound to entail that the actual world is a zombie-like world and that all presently existing living beings, including humans, are unconscious beings. Positions of this kind, as I have already suggested in previous chapters, can be maintained in thought-experiments, but ought not to enter into a realistic attempt to characterise the natural world.

A second and more canonical way to argue in favour of (¬ A) could be to hold that all species feel the same. However, I do not see how one could claim such a thing without encountering serious trouble. Species *have to* differ in how they can feel given their differences in constitution. You cannot feel "redness" if you do not have a neurophysiological apparatus allowing you to have that sort of perceptual experience (i.e., if you do not have some sort of visual apparatus). It follows that all the species lacking the neurophysiological apparatus required for colour vision cannot feel colours, exactly as humans cannot feel by means of eco-location since they lack the physiological apparatus required for eco-location. Thus, bats and human, admitting they both feel *simpliciter*, they do so in a qualitatively different manner. It is goes without saying that the underlying neurophysiological differences separating a human being from the last universal common ancestor are such that the gap of qualitative differences would be even more consistent than that between human beings and bats. The neurophysiological visual apparatus has developed (in many different variants) during the process of evolution by natural selection, but there is no reason to suppose that it was there since the beginning of evolution. More in general, there is no reason to think that *any* of the neurophysiological apparatuses allowing specific kinds of feeling were present at every time since the beginning of evolutionary processes. In fact, this sort of apparatuses require a degree of neurophysiological complexity that even most species living today lack. Since the last universal common ancestor was much simpler than most living beings presently existing – even admitting that it could have had feelings – there is no reason whatsoever to think that it could have had those feelings that require an advanced and fully functioning neurophysiological architecture. Since the global quale of any feeling depends on one's neurophysiological affordances, and since the shaping of neurophysiology is subject to phylogenetic evolution, there are reasons to hold that feeling is at least indirectly subject to qualitative phylogenetic evolution.

Concluding, even admitting that our ancestor species S_1 were feeling species, since they were substantially neurophysiologically different from most presently living species S_2, and since that would have limited the kind of feelings that they could have had, S_1 could not have had the same set of global qualia X that – instead – some species S_2 presently can have. This proves that at least the *qualitative changes claim* about the phylogenetic evolution of feeling is correct: feeling is (at least indirectly) subject to qualitative phylogenetic evolution. I think we can omit highlighting that feelings are "indirectly" subject to phylogenetic evolution because – if we regard feelings as any other biological property such as digestion – it is obvious that the phenomena in question can only evolve (and indeed exist) as long as the physical/functional substrate on which they depend is present. Thus, for the sake of simplicity, I will adopt the claim "feelings are subject to phylogenetic evolution" interchangeably with the claim "feelings are (at least indirectly) subject to phylogenetic evolution".

I have backed up the first (phylogenetic) sense in which we can argue that feeling is subject to diachronic change. For the sake of completeness, before moving to the treatment of the Natural Problem, I go on to consider claim (B), saying that feeling is subject to ontogenetic development[120].

5.4.2 Feeling Is Subject to Ontogenetic Development (b)

The *radical change claim* about the ontogenetic development of feeling says that an individual being at an early developmental stage (B_1) is unconscious, whereas the same individual at a later developmental stage (B_2) can feel (i.e., it is intransitively phenomenally conscious). Is this the case?

The positive answer, (B), says that – ontogenetically speaking – a given biologically individuated persisting individual living being is not a feeling being throughout its developmental history. When a human being is conceived, for example, it does not feel anything at all, whereas babies, children and adults all feel (even though they might feel differently at different developmental stages).

The negative answer, ($\neg$ B), says that feeling is not something that is radically acquired or developed at a given stage of a being's life; rather, it is always present. An individual feeling being has never been an unconscious being in his past ontogenetic history.

The agnostic answer, (B ∨ $\neg$ B), says that we (i) do not know and/or (ii) cannot know whether feelings are something that are submitted to a radical ontogenetic development. We cannot argue for either (B) or ($\neg$ B) because doing so requires taking a stance on whether feelings are or are not subject to a radical ontogenetic development – which we cannot know. As I noted before, agnostic positions such as this can hardly be defeated on epistemological ground, but they are not an option in answering ontological questions. In our world, either it is the case that (B), or it is the case that ($\neg$ B), regardless of whether we do or can know which one is the case.

A first reason to argue for (B) – that feelings are subject to an ontogenetic development – is based on personal experience. I have no memory of feeling when I am not yet born, or as a newborn baby. The amount and variety of feelings I can remember and refer to grow exponentially alongside my maturation as a child and then as an adult. Of course, this could be solely a consequence of the

---

[120] The treatment of this claim is neither directly relevant to the Natural Problem nor required for the understanding of the last chapters. It is possible to move directly to a summary of the findings in section 5.5 and go back to section 5.4.2 at will.

limitation of memory to those late developmental stages, rather than an actual lack of feelings at earlier times. Indeed I think it is highly plausible that we start to feel long before we can remember it, given that – as I suggested in section 4.4 – it might be that we start feeling before becoming adult-like self-conscious and being capable of rationally reflecting on our own conscious experiences. Still, I believe, feeling is constrained in the same way as memory is constrained. We start to remember only when the physiological apparatus allowing us to memorise and recall memories is sufficiently developed. After all – to use a crude comparison – you cannot put any bit of information on a hard disk before all the circuits are physically rightly in place. Similarly, I think, we could be lead to suggest that we start to feel only when the neurophysiological architecture allowing feeling *simpliciter* is sufficiently developed and functioning.

Let us focus on human development. Is feeling *simpliciter* an "all or nothing" phenomenon that is subject to a radical change during ontogenetic development? I think that it is fairly sensible to hold that any embryo at a very early developmental stage (just after fertilization) does not have the minimal neurophysiological structure required to have some (any) feelings[121]. This follows from the premise that a being can only feel as long as it has physically developed sufficiently (in the right way) to make feeling possible, and from the fact that the being in question seems to fail to meet the required standards. As soon as the right sort of minimal neurophysiological and functional architectural standard – whatever that is – is achieved, we begin feeling[122]. If this is right, it would be correct to conclude that we radically start to feel only at a later developmental stage, regardless of whether we can or cannot remember it.

One objection to the claim that feeling is subject to radical ontogenetic development would consist in saying that this mistakenly suggests that feeling

---

121 The topic is delicate and this claim is potentially problematic for ethical reasons, but I believe that the priority has to be to try understanding what actually is the case in the natural world, *not* what should be the case to fit some ethical expectations.

122 I am being forcefully evasive as to what is the "right sort" of minimal neurophysiological functional architectural standard for feeling. This is not due to a lack of will, but rather to a lack of knowledge. Furthermore, in order to make my point I do not need to fill in the specifics; I can just presuppose that there is a matter of fact, even if I concede I ignore what that is. Explaining *what* is the case is the task that NCC research seems to be interested in (see section 1.3), and it is what raises the Hard Problem of consciousness (section 1.4). There is some agreement that the minimal neurophysiological criterion for a being to feel is not related to the mere size or complexity of the neural system, but rather to the way in which the system operates, especially in terms of integration of information (see for example the Global Workspace Hypothesis in Dehaene, 2014). Hopefully someone coming from a department other than philosophy will come up with a convincing story about this.

is something that does not exist for some time and then, at some point in our ontogenetic development, suddenly and mysteriously emerges. But how does this happen? How could a being undergo a radical transition from being a non-feeling being (i.e., a being that cannot feel) to becoming a feeling being? One may think that the transition from a non-feeling being to a feeling being is ontologically puzzling on the grounds that it seems to presuppose a radical shift in the biological identity of the individual. How could a single genetically-individuated being undergo a radical transformation from having an exclusively objective ontology (from consisting only of physical matter) to having a subjective ontology in addition to it? We might say that – biologically speaking – myself at the present time and the foetus that *I* used to be are essentially the very same being at different times because we are genetically identical. Having a given genome is sufficient to grant living beings a biological identity through time regardless of other radical changes to that being. There are countless radical functional and physiological differences between myself now and the foetus that developed to become what I am now. For example, none of the cells that made up the foetus are still alive, the foetus could not digest whereas I can, and the foetus – at least in the first weeks – could not see, whereas I can. I think that holding that my foetus at an early stage did not feel whereas I now do is a similar kind of change that does not undermine the claim that we are always talking about the same genetically-individuated being at different stages in its ontogenetic development. In other terms, the transition from being unconscious to becoming intransitively phenomenally conscious – or from having only an objective ontology to having also a subjective ontology – does not imply a radical shift in the biological identity of a being. Thus, at least in this sense, I see no reason why the radical emergence of an ontologically subjective biological phenomenon, similar to the radical emergence of ontologically objective biological phenomena, could not be a viable option.

Of course, explaining why or how subjective ontology emerges from matter is the really difficult task. If one wants to defend the radical change claim (B) about the ontogenetic development of feeling one has to carry the burden of suggesting how (under which condition) subjectivity can "arise" *ex novo* at a given moment during our neurophysiological development. Just saying that feeling begins to exist when "the right" neurophysiological conditions are present is in fact not explanatory. However, explaining what happens in more detail is a problematic task. Thus the radical change claim about the ontogenetic development of feeling has to come to terms with a rather strong objection. Following this line of thought, feeling is not a characteristic that is radically acquired from scratch at a late stage of a being's life, but rather something that is present since the beginning of our development. This raises an interesting point. There is no way we can know whether an embryo really does not feel anything at all: it might be that it feels something, and that what substantially changes between myself as an embryo and myself in the present is not the *fact* that I feel, but rather the *quality* of feelings I can have.

But is the alternative (¬ B) really better? If there actually is no radical change in the *fact* that beings feel, this entails that even very simple beings such as embryos, however minimally, feel something. But if embryos feel something, by the same token, there is reason to think that even the gametes fusing together to produce a new embryo feel something. In fact, if – say – neither an ovum nor a sperm feel anything at all, on what basis could we suggest that their union feels? How can the fecundation process give rise to feelings on the basis of non-feeling matter? This objection is essentially the same as the one against the radical change claim, only shifted onto an earlier stage in ontogenetic history. How far down the scale should we go in our ontogenetic history to explain the current presence of feeling in developed beings? If we deny the radical change hypothesis, we might have to embrace some form of panpsychism according to which anything that is involved in the ontogenesis of a feeling being feels, and this is a rather unappealing solution (see section 3.2.2). Thus, all things considered, it is not clear that (¬ B) is a better alternative than (B).

Given that the radical change claim does not allow to clearly determine whether feeling is subject to ontogenetic development, I adopt the second test and question whether at least the *quality* in which an individual being feels differs at different developmental stages. The *qualitative change claim* about the ontogenetic development of phenomenal consciousness says that a being S_1 at an early developmental stage is *somehow* phenomenally conscious but cannot have global qualia X, whereas the same being S_2 at a later developmental stage can have global qualia X. The *manner* in which a being is phenomenally conscious is subject to ontogenetic development. Is this the case?

The positive answer, (B), says that – ontogenetically speaking – a given individual being is not an equally feeling being throughout its developmental history. For example, it might be that foetuses in their early stages of development do somehow feel, but cannot feel in the same way as babies, children and adults feel (and not just because of the difference in external stimuli). As I said before, I believe that the quality of feelings a being can have depends on that being's neurophysiological architecture. Since the neurophysiological architecture of a foetus is radically different from that of an adult I think it is sound to suppose that a foetus – if it feels at all – feels in a way that is totally different to the way an adult feels. An adult and a child, on the other hand, have a rather similar neurophysiological architecture and it is therefore imaginable that they could feel in a relatively similar manner. An example should suffice to make this clear. A foetus at an early stage of development cannot feel any sort of taste, not only because it cannot eat, but also because it has not yet developed the gustative apparatus (and without it no gustative feeling is possible). Admitting that feeling capacities depend on physiological development, there are thus reasons to hold that feeling is subject (at least indirectly) to a qualitative ontogenetic development. This view does not assume the radical change claim following which there is a radical shift from not having any feeling at all to having feelings. It just says – in a much more conservative way –, that the quality of the feelings that a being can have during his life varies diachronically in accordance with the being's neurophysiological development.

In order to deny the qualitative change claim, one would have to adopt a version of (¬ B) saying that feeling does not vary qualitatively during a being's ontogenetic development. In other terms, one would have to hold that a being always feels in the same manner throughout his life. However, saying that a being's feelings are not subject to qualitative change during this being's life is a hardly defendable position. I suggest endorsing a "progressive qualitative feeling development" view suggesting that, regardless of whether feelings are (or are not) radically acquired *ex novo* at some point during a being's life, they *are* subject to qualitative ontogenetic development. Someone holding (¬ B) would have to carry the burden of giving a plausible explanation to why and how a being in its early developmental stages could feel (at least potentially) exactly like the very same being at a later developmental stage, despite important neurophysiological differences between the two. I believe that such a burden is too heavy to carry.

Granting that feeling is subject to qualitative ontogenetic development we come to an interesting epistemological remark – call it the "own states of mind problem". We cannot know introspectively what feeling was like (for us) at any of our earlier developmental stages because we are bound to our *present* phenomenological perspective. This is somewhat similar to the other minds problem (see Hyslop 2016), the only difference being that instead of synchronically considering the minds of several individuals, we diachronically consider several states of mind of a single individual. The classical other minds problem says that a being S_1 cannot know whether and how another being S_2 feels like – and *vice versa* –, because every being is bound to his own physically constrained ontologically subjective perspective. Similarly, the "own states of mind problem" says that a being P at a developmental stage t_1 cannot know whether and how himself at another developmental stage t_2 felt like – and *vice versa* –, because each of its temporal constituents is bound to its own present physically constrained ontologically subjective perspective. The being you are now is distinct in feeling capacities from the being you were twenty years ago, or at your birth, or prior to your birth. This qualitative distinctness of all the historical "you" making up your developmental history causes a problem of accessibility similar to that of other minds: at the present time you can only know how it feels like to be yourself now. But, *now,* you cannot subjectively know how it felt like to be you at the time of your birth, or when you were ten years old. Only then, at those moments respectively, could you have known what it felt to be you in that moment. In other terms, since how we feel changes in time with our physiological development, and since we can feel nothing but what our present neurophysiological architecture allows us to feel, there are good reasons to be skeptic with regard to the attribution of specific global qualia to ourselves at previous developmental stages on the basis of introspection. This epistemic constraint, suggesting that we cannot really know how it felt like to be ourselves at previous developmental stages, could eventually lead one to question whether it is really the case that feeling is subject to qualitative ontogenetic development, since – after all – we cannot know anything but how we presently feel. Again, however, I think that if we want to make genuine progress in better understanding consciousness we ought to be pragmatic and realistic, and this sort of skepticism is counterproductive for such a goal. I suggest putting it aside, and sticking with the progressive qualitative changes claim suggesting that feeling is (at least indirectly) subject to a progressive qualitative ontogenetic development.

5.5 Summary of the Claims and of the Conclusions

In section 5.4 I have argued for the following claims:
A) Feeling is (at least indirectly) subject to qualitative phylogenetic evolution.
B) Feeling is (at least indirectly) subject to qualitative ontogenetic development.

By arguing that feeling is (at least indirectly) subject to qualitative phylogenetic evolution (section 5.4.1) and that feeling presently exists (section 5.1), I have justified the assumption of such claims in the formulation of the Natural Problem (section 3.3). This was the primary aim of the chapter. The secondary aim of this chapter was to show that there are reasons to think that, in the actual world, phenomenal consciousness is subject to diachronic change both phylogenetically and ontogenetically. I have done so on the basis of a long and hopefully fruitful taxonomical work, by individuating and individually supporting the diachronic claims (A) and (B). Defending these claims has also hopefully highlighted the vastness and interrelatedness of interesting consciousness-related questions that can arise if we adopt a diachronic framework.

I now lay the questions about ontogenetic development of feeling aside, and focus the remaining chapters on trying to advance an answer to the Natural Problem of explaining why feeling presently exists.

6 Why Do We Feel?

In the last chapters I have clarified what I mean by the expression "being conscious" (chapter 4) and supported the claim that feeling presently exists and is (at least indirectly) subject to qualitative phylogenetic evolution (chapter 5). On this basis I can now introduce the theory of evolution by natural selection (section 6.1.1) and outline the strategy I intend to pursue (section 6.1.2) to develop a hypothesis suggesting *why* feeling presently exists – a possible reply to the Natural Problem[123]. In order to clarify such a strategy, I distinguish two notions of function (section 6.2) and separately sketch the possible answers to the question of knowing whether feeling has an "evolutionary" function (section 6.3) and a "biological" function (section 6.4). This prepares the ground to outline my hypothesis in chapter 7.

6.1 Feeling in an Evolutionary Framework

6.1.1 A Darwinian Evolutionary Theory

In evolutionary biology, following a standard definition, a population is said to evolve when a new gene is introduced, an old one disappears, or when the mix of genes is altered. Following this definition, when there is change in the gene frequencies (computed *per capita*) found in a population, evolution occurs[124]. Evolutionary biology, however, considers as evolution also cases of *phenotypic change* – typically changes in morphology, physiology, and behaviour – that are due to change in the genetic endowment of a given population. I believe that feeling is subject to phylogenetic evolution in the latter sense. That is, I take feeling to be a phenotypic characteristic that has been acquired (in the radical change hypothesis) or has changed (in the qualitative change hypothesis) due to change in the genetic endowment of some members of a given population.

---

[123] It should be clear since the beginning that what follows is a philosophical *theory* about feeling in an evolutionary setting, and that as such (as any theory) it involves some degree of speculation and conjecture. Since the subject matter (consciousness) is an ontologically subjective one, the amount of speculation and conjecture is higher than in most scientific theories, and indeed this is the reason why science has for a long time ignored the subject. Keeping this in mind is important, but it ought not stop us from trying to put forward a reasonable hypothesis about why feeling presently exists.

[124] There are reasons to think that gene frequency is sufficient but not necessary to explain evolution, i.e., evolution does not require change in gene frequency. See (Sober 2000, pp. 2–5).

Before I can tackle the questions of *why* these specific phenotypic characteristics have been acquired or have changed in the process of evolution, I need to say something more about the framework I intend to use for my argument, namely Darwin's theory of evolution by natural selection. Most of it is likely to sound familiar, since I will not go to greater detail than needed for the present scope, but anyway *repetita iuvant*.

Current evolutionary theories about how life evolves all take Darwin's paradigm as a starting point. Darwin's main contribution consisted in combining and applying two already existing ideas to build a comprehensive theory of evolution (for an introduction see, Sober 2000). The first idea is the strong idea that all present and past terrestrial species are genealogically related and form a *single* tree of terrestrial life[125]. According to this idea, for any two current species that now populate the earth there is a species that is their common ancestor. Moreover, all contemporary species descend from a common universal ancestor. Of course, if human beings and chimps (or human beings and any other specie) have a common ancestor, the lineages leading from the ancestor to its descendants must have undergone some modification. Taken alone, however, the idea of different species being united into a single tree of life does not explain *why* modifications in the composition of populations occur. Why do novel characteristics evolve in a lineage? Why do new species come into existence (i.e., why are there speciation processes)? Why do some species become extinct? In sum, what are the causes of evolution?

Darwin's second idea, natural selection, intervenes to answer these questions. As Sober (2000, p. 11) puts it, natural selection is the principal explanation of *why* evolution has produced the diversity of life forms we observe[126]. We can distinguish two main sorts of evolution – microevolution and macroevolu-

---

[125] Compare with Lamarck's conception of evolution following which present and past species do not form a *single* tree of life. For Lamarck, living beings contain an inherent tendency to increase in complexity, and they do so by means of descent with modification. Humans, the most complicated of creatures for him, descend from complicated species, which in turn descend from simple life forms, which in turn descend from non-living material. Simple creatures belong to a younger lineage than complex creatures, and this explains why – following Lamarck – present-day humans and present-day earthworms are not related. For him, present-day human beings descend from earthworms that lived long ago.

[126] I focus on the process of natural selection, but there are other causes of evolution. Change in gene frequency can also be caused by mutation (producing new single genes), migration, or random genetic drift (Kimura 1983). Moreover, evolution can be driven by recombination, a process that alters the frequency of combinations of genes (i.e., it produces new combinations) thereby enriching the range of variation. All these causes, including natural selection, might contribute individually or simultaneously at the process of evolution.

tion – and they both can be explained by natural selection. I begin by explaining how natural selection can explain microevolution, that is, those modifications that take place within given persisting existing species. I then do the same with respect to macroevolution.

Consider the following points presenting a case of microevolution in the phenotypic characteristics of a species:
(i) There is a variation in the phenotypic characteristics of the members of a given population (variation can have a variety of causes). *E.g., some members of a zebra population run at a higher speed than others (variation of running speed).*
(ii) This variation entails a variation in fitness, i.e., a variation in the ability to survive and reproduce of the members of a given population. *E.g., zebras with a high running speed are better able to avoid predators and thus to survive than slower zebras (i.e., the variation of running speed makes a difference to the survival or reproduction of zebras).*
(iii) Because of genetic similarity, the offspring tend to have the same phenotypic characteristics of their parents, i.e., offspring phenotype is inherited from parental phenotype[127]. *E.g., on average, zebra offspring tend to resemble their parents with regard to running speed (the running speed is inherited). Thus, offspring of fast parents tend to be faster than the offspring of slow parents.*

Following Darwin, if conditions (i), (ii), and (iii) are in place, then zebras with a high speed as phenotypic characteristic will be favoured by natural selection over slower zebras. This is a case of microevolution because gradually, over many generations, the average running speed in the herd of zebras (a single persisting species) will change by tending to increase. The above example shows how natural selection can intervene to modify some phenotypic characteristics of a given lineage. In sum, the process of natural selection in microevolution operates when there is a *heritable variation in the fitness* in some members of a given population (Lewontin 1970).

---

127 Notice that the tendency of offspring to resemble their parents in some of their phenotypic characteristics does not imply that the genes *always* play a role in shaping these similarities. As Sober (2000, p. 11) rightly points out, genetic similarity between offspring and parents is an obvious explanation for the heritability of phenotypic traits, but it is by no means the *only* possible explanation. For example, the heritability of a high running speed could be explained by the fact that fast zebra offspring have the same (rich) dietary regime of their fast parents, whereas slower offspring have the same (poor) dietary regime of their slow parents.

It might be that the diachronic changes in feeling occur within a given persisting existing species, and that therefore they count as a case of microevolution such as the one presented here. Consider the following example:

(i) There is a variation in the phenotypic characteristics of the members of a given population (variation can have a variety of causes). *E.g., some members of a population feel differently than others (qualitative change in feeling).*

(ii) This variation entails a variation in fitness, i.e., a variation in the ability to survive and reproduce of the members of a given population. *E.g., appropriately feeling beings are better able to survive than inappropriately feeling beings*[128] *(i.e., the variation in how one feels makes a difference to survival or reproduction).*

(iii) Because of genetic similarity, the offspring tend to have the same phenotypic characteristics of their parents, i.e., offspring phenotype is inherited from parental phenotype. *E.g., on average, offspring of appropriately feeling beings tend to resemble their parents with regard to feeling (feeling appropriately is inherited). Thus, offspring of appropriately feeling parents tend to feel more appropriately than offspring of inappropriately feeling parents.*

As in the zebra case above, if conditions (i), (ii), and (iii) are in place, then beings with appropriate feelings as phenotypic characteristic will be favoured by natural selection over inappropriately feeling beings. This is a case of microevolution because gradually, over many generations, the average quality and frequency of appropriate feeling in the population under scrutiny (a single persisting species) will change by tending to increase.

Besides explaining microevolution, Darwin's idea of natural selection can also explain macroevolution. More precisely, it can explain both extinction, the process of disappearance of old species and higher taxa, and speciation, the process of origination of new species and higher taxa. Both extinction and speciation can have different causes, and they both entail a change in gene frequency.

Modern evolutionists distinguish two different variants of speciation processes. The first one is *analgesis*, a process by which small evolutionary modifications within a single lineage (i.e., microevolutions within a single specie) add up gradually, reaching a point when ancestors and descendants are so different that they should be viewed as different species. Analgesis is thus a specific variant of speciation process by which a new species originates in a 1:1 proportion from and

---

[128] What I *mean* by "appropriately" and "inappropriately" feeling will be explicated later. For the time being it is sufficient to suppose that "appropriately" feeling is qualitatively different and biologically more efficient than "inappropriately" feeling.

old species of the same lineage. This also entails a process of *extinction*, since the old species becomes extinct and is substituted by a new one.

There are no uncontroversial criteria to discriminate one species from the other. One standard criteria, though, consists in saying that two populations belong to different species if they are reproductively isolated from each other, i.e., if they cannot produce viable fertile offspring with each other (see Sober 2000, p. 14). Thus, if it is the case that ancestors and descendants coming from a single lineage become reproductively isolated, they are viewed as members of different species, and therefore subject to a process of analgesis.

Fast zebras and their slow zebra ancestors belong to the same species (i.e., they are *not* subject to analgesis) because the variation in speed alone (however important) does not alter the ability of fast and slow zebra to reproduce with each other. Is it possible that small evolutionary modifications in feeling capacities within a single lineage (i.e., microevolutions in feeling within a single specie) added up gradually reaching a point where ancestors and descendants were reproductively isolated from each other, so as to be viewed as different species? I believe it is hard to imagine *why* any variation in feeling alone – that is, other things being equal – would alter the *ability* to reproduce with differently feeling beings within the same lineage. I think that, assuming that philosophical zombies exist and belong to the same lineage as humans, humans and zombies would have to count as members of the same species, since by definition they are functionally identical and this entails also that they could produce viable fertile offspring with each other. Their phenotypic differences in feeling, like the difference in speed in zebras, would not be a sufficient criterion to argue that humans and zombies belong to different species. In short, I believe that change in feeling alone cannot induce a process of analgesis, be it a radical change or a qualitative change[129].

A second sort of process that can explain speciation is *cladogenesis*, a (synchronic) branching process by which similar organisms living in different environments come to diverge from one another because of how natural selection intervenes. Cladogenesis explains how, depending on environmental conditions, a single species can give rise to more than one descendant species (in a one-to-

---

[129] Marcel Weber (personal communication) commented that it seems that emotions *could* make a difference in mate selection. I agree that feeling (including having emotions) could play a role in the *actualisation* of reproductive behaviour. However the point here is that the presence of feeling is arguably not *necessary* for two individuals to technically reproduce, since the fecundation process itself is purely mechanical. If this were not the case *in vitro* fecundation would not be possible, since there are no emotions directly involved in this process.

many proportion). Again, we can assume the standard criteria saying that two populations belong to different species if they are reproductively isolated from each other, i.e., if they cannot produce viable fertile offspring with each other. In virtue of processes of cladogenesis we can explain why the number of species tends to increase in time, as shown by the ramifications of the tree of life, rather than to remain constant (as it would be if analgesis were the only kind of speciation process).

Could a variation in feeling play a role in a process of cladogenesis? I believe it could. If feeling beings from the same specie live in radically different environments, they might come to diverge from one another because of how natural selection intervenes. Beings having inappropriate feeling capacities might easily manage to survive in simple environments, but not manage to survive in ever-changing complex environments. If this were correct, the individuals living in simple environments would not undergo a selective pressure linked to feeling capacities, whereas those living in the complex environment would. Eventually, because of that, the two populations might come to diverge to such an extent that would make them into separate species[130].

Answering the above questions about the role of feeling in specific sorts of evolutionary processes such as analgesis or cladogenesis is not required to face the Natural Problem[131]. What matters for the present analysis is simply that natural selection is a good candidate to explain *why* evolution has produced the diversity of life forms we observe in the tree of life, regardless of the specific processes involved[132].

I assume that the tree of life hypothesis and the theory of evolution by natural selection correctly account for actual biological evolution, and I base my hypothesis geared toward the Natural Problem on these foundations. I think this move is justified by the fact that both the tree of life hypothesis and the theory of evolution by natural selection are widely accepted in the scientific community (to the

---

130 Marcel Weber (personal communication) argues that in this scenario feeling differences play no causal role in the cladogenesis because the trait divergence here is just a side effect. This however need not be the case. I tackle this question in section 6.4 and especially in chapter 7.

131 However, it is fascinating to see that once we start to look at consciousness from an evolutionary perspective, there are a number of unexpected and interesting philosophical questions and paths of analysis one might want to pursue.

132 Marcel Weber (personal communication) commented that I am presupposing an unnecessary strong adaptationism, given that my claim only holds in the absence of other evolutionary forces (constraints). This is true, but I have to start somewhere, and I think it is fair to concede me this rough approximation given that it simplifies the context and it does not significantly affect my theoretical goals.

extent of even being often – and wrongly – taken for granted in biology) because of their growing solidity, durability in time, and compatibility with empirical evidence. In the scientific world there is a wide reaching and well-spread consensus that Darwin's theory of evolution by natural selection (or its reviewed versions) is the most rational explanation available for the mechanism behind the evolution of biological organisms[133]. Not only is the combination of the tree of life hypothesis and the theory of evolution by natural selection compelling to explain evolution, but it has been corroborated to such an extent in the biological sciences that the burden of proof is on whoever wants to deny these premises in the framework of an evolutionary approach to biological phenomena.

6.1.2 A Sketch of the Hypothesis

In previous chapters I have claimed that in the actual natural world feeling is an ontologically subjective biological phenomenon, rather than some kind of mysterious or God-given extra ingredient. There is a counterfactual reason to hold this: in the actual natural world, the absence of naturally evolved biological beings in a certain context entails the absence of subjectivity in that context[134]. In other words, since feeling is a phenomenon that only a subset of naturally evolved biological beings actually exhibits, it is reasonable to consider it a biological phenomenon.

Since feeling is a biological phenomenon – in accordance with methodological naturalism – it is reasonable to adopt the explanatory tools of biology to analyse it, as for any other natural phenomena such as digestion or photosynthesis (to take examples often cited by John Searle). As we have seen, natural selection tells us that in the long run any feature that increases a being's biolog-

---

[133] I do not engage in an apology for Darwin's theory or in a debate against creationism. Notable scholars such as Richard Dawkins (1976, 1986) did this extensively, convincingly, and much better than I could ever aspire to do.

[134] It might be argued that the absence of naturally evolved biological beings in a certain context only implies the absence of *natural* feeling in that context. In fact, if we were able to engineer – say – feeling robots, we could put one in a room and say that there is (artificial) feeling in that context despite the absence of a natural biological being. However, my point is that it is a contingent fact that feeling is at least originally *only* a natural biological phenomenon, since – I claim – it evolved long before such things as artefacts existed (see section 1.5). I do not think it is *in principle* impossible to create artificially feeling beings in the future, but even if we could to that, at least indirectly, feeling will still have its historical roots and raw model in naturally evolved biological systems.

ical fitness (i.e., the chance of survival and reproduction) in a given environment tends to be selected given that the stronger the biological fitness, the higher the probability that the being in question will manage to pass on his genes to the subsequent generation, thereby preserving what made it biologically "fit". Digestion and photosynthesis were selected because they were useful for some beings. Similarly, – I think – feeling could have been selected because of its usefulness to some beings.

The first coarse-grained answer to the question "why feeling presently exists" is thus straightforward. At some point in evolutionary history, when the right sort of neurophysiological architecture and functioning came into existence, feeling beings started to exist (i.e., some beings started to feel)[135]. In given environmental conditions feeling increased the biological fitness of some beings with respect to other beings. By virtue of natural selection, some feeling beings were selected in the long run, transmitting their phenotypic traits (with constant modifications) to their descendants, and so on up until now. This could explain evolutionarily why there presently are feeling beings.

This might be a sound coarse-grained hypothesis, but it comes too quickly and not without problems. In order for this story to be plausible, feeling has to somehow increase a being's biological fitness, i.e., it has to be causally efficient. But is this the only possible way to explain why feeling presently exist? Moreover, even admitting that feeling can increase biological fitness, the problem is that of understanding *how* the fact of feeling could cause anything at all. More precisely, the problem is that of understanding how the ontologically subjective dimension of feeling (i.e., what defines feeling as an ontologically subjective phenomenon) could cause anything that its ontologically objective neurophysiological supervenience base alone could not cause. I raise these questions in section 6.4 after having introduced an important distinction.

6.2 Distinguishing Kinds of Function

In order to proceed on safe grounds, I need a brief excursus to draw attention to an important distinction between two senses of the term "function". The question that lies at the centre of my investigation of feeling is that of the biological *causal*

---

[135] I take this as a bare fact, but I do not try to explain how consciousness first originated. This is a synchronic Hard Problem requiring explaining the specific subserving mechanisms underlying phenomenally conscious states of mind. The problem I am concerned with – the Natural Problem – is that of explaining why feeling *presently* exists, given that it is subject to phylogenetic evolution.

function of consciousness. Answering this question, however, does not amount to suggesting what the *evolutionary* function of feeling is. Biological causal function and evolutionary function are two conceptually distinct notions[136].

An investigation of the biological *causal* function of feeling aims at determining whether feeling has biological causal powers or causal dispositions, and – if so – what they are. That is, it aims at determining how feeling could contribute to the biological fitness of a phenomenally conscious being. On the other hand, an investigation of the *evolutionary* function of consciousness aims at determining what is the relation between feeling as it presently is and feeling as it originated in the past, trying to determine if feeling, as it presently is, was selected for.

We can outline the argument leading to the question of whether feeling has a biological causal function (i.e., whether it can contribute to the biological fitness of a given being) as follows:

1) At any time in phylogenetic evolution either (i) feeling has a biological causal function or (ii) feeling does not have a biological causal function – the two being mutually exclusive.
2) Feeling has a biological causal function if and only if it positively contributes to the biological fitness of the members of a given species.
3) Could feeling have a biological causal function? That is, could feeling positively contribute to the biological fitness of the members of a given species? How?

We can outline a different argument leading to the question of whether feeling has an evolutionary function as follows:

1) Feeling presently has a biological causal function.
2) Feeling has an evolutionary function if and only if it was selected for its present biological causal function (i.e., if it is an adaptation).
3) Does feeling have an evolutionary function? That is, was feeling selected for the biological causal function it presently has?

---

[136] Notice also that the way in which I use the term "function" is neutral with respect to its use in etiological or dispositional theory. In etiological theory the function of a trait is what that trait did in the past and was selected for. In systemic (or dispositional) theory, the function of a trait is its present disposition to be adaptive. The Natural Problem requires only explaining the past causal function of feeling – how feeling could have been adaptive in the past. For feeling to have been adaptive in the past it must have manifested its dispositions in the past. Thus, I can tackle the Natural Problem without having to take a stance in the debate about how the *current* function of feeling is to be individuated. I thank Karen Neander for pointing this out to me.

These two arguments suggest that the question of the biological causal function is distinct from that of the evolutionary function, even though the two are somewhat connected. The connection lies in the fact that in order to answer the question about the evolutionary function of consciousness, one has first to determine whether feeling presently has a biological causal function. That is, one cannot give an answer to the question regarding the evolutionary function of feeling without first answering the question about its biological causal function[137]. Knowing whether (and how) feeling has a biological causal function is a necessary (even if not sufficient) condition to know whether feeling has an evolutionary function. This suggests that the question of the biological causal function of consciousness is more fundamental in the order of explanation than the question of its evolutionary function.

I believe that a comprehensive diachronic theory of consciousness should say something about both the biological causal function and the evolutionary function of feeling. I briefly summarize the possible replies to the question of knowing whether feeling has an *evolutionary* function (6.3). However, since I am mainly concerned with answering the Natural Problem – and since anyway giving an answer to the question about the evolutionary function requires explaining whether feeling could have a biological causal function and what it could be – I then leave the evolutionary function question aside and focus only on the more urgent and crucial question of knowing whether feeling could increase biological fitness (section 6.4).

6.3 The Evolutionary Function of Feeling

I consider feeling as a phenotypic feature currently present in some living beings in the actual world. How can we characterise it in terms of its evolutionary function? There are several options. I summarize all the possibilities in Table 7 and I separately explain the different possibilities[138].

---

[137] Notice that, on the contrary, one can in principle answer the question about whether feeling has a biological causal function (and what it is), without having to take a stance on the question about whether feeling has an evolutionary function.

[138] I borrow the relevant terminological distinctions that I use to talk about feelings in an evolutionary perspective from Stephen Gould (Gould & Lewontin 1979; Gould & Vrba 1982; Gourba 1997).

Table 7

		Trait's original primary function (genesis):		
Feeling in an evolutionary perspective		Aptation with function X	Random nonadaptive by-product of another trait (no adaptive function)	Necessary (consequent) nonadaptive by-product of another trait (no adaptive function)
Trait's current primary function (fitness):	fit for X	**Adaptation** (evolved for X, it presently has adaptive function X)	**Exaptation** (evolved randomly, it presently has non-adaptive effects X)	**Spandrel** (evolved as necessary consequence of trait Z, it presently has non-adaptive effects X)
	fit for X and potentially fit for Y	**Preaptation** (evolved for X, it presently is a potential exaptation with effects Y)	**Preaptation** (evolved randomly, it presently has non-adaptive effects X, and it is a potential exaptation with effects Y)	**Spandrel** (evolved as necessary consequence of trait Z, it presently has non-adaptive effects X, and it is a potential exaptation with effects Y)
	fit for Y	**Exaptation** (evolved for X, it presently has random non-adaptive effects Y)	**Exaptation** (evolved randomly, it presently has non-adaptive effects Y)	**Spandrel** (evolved as necessary consequence of trait Z, it presently has non-adaptive effects Y)
	not contributing to fitness	**Nonaptation** (evolved for X, it presently has no effects)	**Nonaptation** (evolved randomly, it presently has no effects)	**Nonaptation** (evolved as necessary consequence of trait Z, it presently has no effects)
	negatively contributing to fitness	**Maladaptative Trait** (evolved for X, but presently X is maladaptive)	**Maladaptative Exaptation** (evolved randomly, it presently is maladaptive)	**Maladaptative Spandrel** (evolved as necessary consequence of trait Z, it presently is maladaptive)

6.3.1 Feeling as Nonadaptation

Nonadaptations ("nonaptations", in Gould & Vrba's terminology) are phenotypic features not presently contributing to biological fitness, i.e., not modifying in any way a being's chances of survival and reproduction. If feeling does not currently have any causal function it is a nonadaptation. However, saying that feeling is presently a nonadaptation does not imply that it has always been so and always will be so. It might be that feeling did originally have a causal function, or that it will acquire a causal function in the future. If feeling is an epiphenomenon, then it is a nonadaptation, since by definition epiphenomena cannot cause anything at all. Because of economy, nonadaptive traits are often not selected against. This might explain why we still feel even if feeling does not play any causal role.

6.3.2 Feeling as Maladaptive Trait

Maladaptations are traits that, in a given context, are more harmful than they are useful to the organism. This term only suggests "in which way" a trait contributes to a being's overall biological fitness by specifying that it decreases it. A trait can suddenly become maladaptive due to a radical change in environmental conditions that alters the way in which that trait contributes to the fitness of a given being. Maladaptive traits are often selected against in the long run. Feeling is a maladaptation if, regardless of its evolutionary function, its current causal function negatively contributes to the being's overall fitness. If this were the case, it might be that feeling will be selected against in the future and disappear.

6.3.3 Feeling as Adaptation

Adaptations, as Gould & Vrba define them (1982, p. 6), are features that positively promote fitness and – in addition to that – were built by selection for their current role. Thus, determining whether feeling is an adaptation depends on the criterion of historical genesis. Even if feeling currently has a causal function, it does not follow that it is an adaptation in evolutionary terms. It is an adaptation if and only if, in addition to currently having a causal function, it was built by selection *for* that role. The terminological trap lays in the fact that feeling has an *evolutionary* function if and only if it was selected *for* its present role (i.e., if it is an adaptation). In fact, only the operation of an adaptation built by selection for its current role is called 'function' in evolutionary terms; the operation of anything else is instead called 'effect'.

6.3.4 Feeling as Exaptation

Exaptations are features evolved for other usages than the present one (or for no function at all) and only later co-opted for their current role. As Gould & Vrba put it, exaptations are features that are *"fit (aptus) by reason of (ex) their form, or ex aptus"* (1982, p. 6, original italics). In other terms, they are features that owe their present fitness not to their adaptive advantage in the past, but rather to the fact of having been co-opted at a later stage in evolutionary history. Environmental conditions play a major role in determining whether a by-product of another trait comes to accidentally increase the overall fitness of the being, accidentally decreases it, or does not contribute to its fitness at all. Thus the very same non-adaptive trait could be co-opted in given conditions, but abandoned in other ones. Feeling may have originated as an evolutionary accident, as a by-product of neurophysiological evolution, and be an exaptation if, despite currently having some causal function, it was not selected *for* it (i.e., if it was not originally selected *by virtue* of its present causal function). In this case, its current contribution to fitness is not a direct consequence of adaptation. Remember that, in the evolutionary perspective, only adaptations have (adaptive) functions. The operation of an exaptation (i.e., a feature not built by selection for its current role) is therefore called 'effect', rather than 'function'. So if feeling is an exaptation and has causal efficacy it would be wise to say that it currently has causal *effects*, rather than causal *functions*, in order to avoid misunderstandings.

6.3.5 Feeling as Preadaptation

Preadaptations (or "preaptation", as Gould & Vrba 1982 suggest) are potential, but unrealised, exaptations. They are "*aptus* – or fit – before their actual cooptation". That is, they are a category of exaptations considered before they originate (randomly) with respect to their new effects. For example, thinking retrospectively about the feathers encasing the running of dinosaurian ancestors of birds, we can describe them as potential exaptations for flight (or preaptations).

6.3.6 Feeling as Spandrel

Spandrels are phenotypic features that (i) are not a direct product of adaptive selection, and (ii) are necessary by-products of the evolution of some other characteristic (see Gould & Vrba 1982). Notice that (i) is the same criteria for exaptations, but (ii) makes a stronger requirement in the modality in which the spandrel is linked to the selected trait (i.e., necessity). Spandrels are the result of architectural requirements (constraints) in the *Bauplan* of an organism. The architectural term spandrel was borrowed by Gould & Lewontin (1979) to designate those forms and spaces that arise as necessary by-products of another decision in design, and not as adaptations for direct utility in themselves. The classical example from architecture is that of the spandrels of San Marco, and the claim was that even though they might have some causal function now, it does not follow that spandrels had been built to fit that function (i.e., that the causal function they have currently is their adaptive function). Actually, in this case, it is obvious that spandrels are a necessary architectural consequence of building two arches one next to the other, i.e. if you build that you will have spandrels as a necessary by-product. Feeling is a spandrel if – regardless of its current causal function or absence thereof – it is (individually) a necessary by-product of a physiological system with given characteristics (say – a central nervous system working in a specific way).

6.3.7 Conclusion on the Evolutionary Function of Feeling

As things stand, there are no conclusive reasons to hold that feeling is – say – an adaptation, an exaptation or a spandrel. For the scope of the present research, determining the precise evolutionary function of feeling would be by far a too ambitious project. In fact, this would not only require knowing what the current biological causal function of feeling is, but also whether feeling was selected for it. If feeling presently has a biological causal function – something we still need to determine –, then in terms of evolutionary function feeling is either an adaptation, an exaptation or a spandrel. On the other hand, if feeling has (and has had, in the past) no causal function whatsoever, then in terms of evolutionary function feeling could be described as a non-adaptive spandrel. All these options are in principle viable, but only one correctly describes the evolutionary function of feeling.

Answering the question of the evolutionary function of consciousness is probably one of the hardest tasks one could think of in the philosophy of mind. Exposing the different possibilities, however, highlights once more how rich and variegated the domain of questions about the nature consciousness becomes once we consider consciousness diachronically and as biological phenomenon.

6.4 Could Feeling Have a Biological (Causal) Function?

In section 6.1.2 I have outlined the strategy I intend to pursue to develop a hypothesis suggesting *why* feeling presently exists – a possible reply to the Natural Problem. The strategy hinges on showing that – and how – feeling could have (or, at least, could have had in the past) a biological causal function. In fact, if feeling could have increased the biological fitness of some species in the past we could then give a reasonable explanation, by means of the theory of evolution by natural selection, to the diachronic question of knowing why feeling presently exists. But could feeling really have a biological causal function? I separately present and examine two opposing views concerning this question. I begin by presenting the view according to which feeling does not have a biological causal function (section 6.4.1). I then suggest some reasons for putting this view aside and to assume instead the alternative view suggesting that feeling has a biological causal function (section 6.4.2).

6.4.1 Epiphenomenalism

Epiphenomenalism, in its general formulation (see section 2.2.2), is the thesis according to which mental states exist, but cannot cause anything (neither mental nor physical states). Applied to the picture of feeling that I have endorsed, epiphenomenalism suggests that feeling does not have a biological causal function. More precisely, it suggests that, even granting that feeling presently exists as an ontologically subjective biological phenomenon, it cannot cause anything and therefore it cannot influence (never mind increase) a being's biological fitness. If one endorses epiphenomenalism it would be possible to answer the Natural Problem by arguing that feeling presently exists by virtue of being a by-product of something else that has a biological causal function. It could be that feeling originated randomly in phylogenetic evolution as a by-product of something else and has been preserved throughout the process of evolution until now just because it came "for free" with some other trait, but not because of its own contribution to the biological fitness of feeling beings[139]. This would by no means be something unique in biology[140].

---

[139] It might be argued that feeling, despite being an *epiphenomenal* by-product of other traits and not having causal powers has nonetheless some hedonistic effects.

[140] There are traits that presently have no use whatsoever (nor, generally, disadvantages) and that have not been dismantled by natural selection just because doing so would cost energy and give no positive contribution to the fitness of the being in question.

I am not particularly annoyed by this version of epiphenomenalism. I think it is a fair logical possibility that ought to be taken seriously. I am not arguing *for* epiphenomenalism here, but I want to point out that maybe it is just the case that feeling is epiphenomenal and that's the end of the story. The only big trouble with epiphenomenalism is that it offers an explanatorily suspicious answer to the Natural Problem. We may in principle be able to solve the Hard Problem and explain what (synchronically) causes epiphenomenal feelings, but it seems suspicious to hold that the existence and persistence of feelings throughout phylogenetic evolution is not due to something that feeling does. It would be certainly surprising if a peculiar, incredibly rich and variegated biological phenomenon such as phenomenal consciousness were just a twist of fate, a useless epiphenomena that nature brought about by chance and never got rid of. However, not necessarily everything in nature makes sense. Feeling might be just an epiphenomenal by-product of evolution. If that were the case, we should consider ourselves blessed with the most incredible and fancy randomly selected accessory one could hope for: ontological subjectivity.

But too much luck – although welcome – should look always suspicious. It might be that feeling does not have causal powers, does not contribute to biological fitness, and that it has been preserved by chance, but we should not assume this is the case before having excluded all the more rationally sound and explanatory options[141]. Arguably the most explanatory answer to the Natural Problem would consist in saying that feeling has (or at least has had) a biological causal function, and that the present existence of feeling can be explained by the fact that feeling has been selected or co-opted in the past by virtue of what it could *do*. Given that this sort of reply would be explanatorily more satisfactory than the epiphenomenalist one, I believe we have good reasons to begin by assuming that something like this is likely to be the case, and try to explain *how* feeling could have a biological causal function. If at the end of the day, after a thorough analysis, no such biological causal function can be identified, I would be the first to retreat into epiphenomenalism about consciousness for the sake of philosophical consistency.

---

[141] Remember that I endorsed the desiderata of Epistemic Force (or Explanatory richness) as the decisive criterion to measure the value of otherwise equally viable competing theories of the mind-body problem (section 2.3). Now I apply the same desiderata to decide between two otherwise equally viable answers to the Natural Problem.

6.4.2 Reasons To Put Epiphenomenalism Aside

Feeling may or may not be epiphenomenal, but there is no decisive *a priori* reason to endorse either one of these options. There are two strategies that one may attempt to sort this controversy out, and they both begin by assuming – by hypothesis – that consciousness is epiphenomenal. The first strategy consists in proving that everything that has a cause is totally caused by physical processes and that there is no causal overdetermination, or that feeling does not cause anything at all. By so doing we would exclude the possibility of something being caused by feeling (i.e., by their ontologically subjective dimension). This strategy, of course, is sound in theory, but unmanageable in practice. The second and more viable strategy consists in trying to *falsify* the hypothesis that feeling is epiphenomenal by (i) focusing on those cases in which feeling seems to cause something, and (ii) verifying if this is the case. One case in which feeling is proved to play a role in influencing biological fitness would be sufficient to dismantle the hypothesis that feeling is epiphenomenal. Again, however, doing so empirically is very difficult, since it is not clear how we could verify whether it really is a feeling (i.e. a phenomenally conscious state of mind) that has caused something, rather than its subserving neurophysiological base.

There are radical attempts to get rid of epiphenomenalism. Nichols and Grantham (2000), for example, adopt a complexity argument and suggest that since phenomenal consciousness is structurally complex, this provides prima facie evidence that phenomenal consciousness is an adaptation. The argument is based upon an idea coming from evolutionary biology, suggesting that the structural complexity of a given organ can provide evidence that the organ is an adaptation, even if we know nothing about the causal role of that organ. Even though feeling is definitely not a physical organ, and even though I believe that the argument by Nichols and Grantham is not strong enough to logically exclude the possibility of feeling being epiphenomenal[142], I think it can be used to at least *support* the intuition that feeling has some causal power. This does not mean I agree with them in saying that phenomenal consciousness is an adaptation. Indeed, as we have seen (section 6.3), it could also be an exaptation or a spandrel. But I do agree that the apparent complexity of feeling – intended as the qualitative richness of phe-

---

142 The problem is that it is not clear what it means for consciousness – an ontologically subjective phenomenon – to be "structurally complex". Structural complexity is easily definable for ontologically objective phenomena, but not so for consciousness. Perhaps the phenomenological structure of global qualia could count as complex, but one would have to carefully argue for this claim.

nomenology – as well as the consistency of its presence in members of the same species, are *indicators* suggesting that the phenomenon of feeling is likely to be the reason for its own existence. In other words, given the apparent complexity and the uniquely ontologically subjective mode of existence of feeling intended as biological phenomenon, it seems unlikely that feeling exists despite not doing anything at all. This of course is not a bulletproof argument, but rather an intuitive reason to disregard epiphenomenalism as an explanatorily satisfactory solution to the Natural Problem.

In this deadlock situation, the best way to discard epiphenomenalism is on the basis of a methodological argument. If epiphenomenalism is right, then there is nothing left to explain about the biological causal function of feeling since feeling simply does not cause anything at all: game over and explanatory ambitions miserably deluded. On the other hand, if epiphenomenalism is wrong, then there is something more left to explain – namely, *how* feeling can influence biological fitness. Thus, even if epiphenomenalism is a logically valid hypothesis, we are not justified in assuming it is true before and without having first excluded the potentially more explanatory alternative. Instead of endorsing epiphenomenalism and resting on the uncomfortable laurels of this explanatorily suspicious solution to the Natural Problem, it makes sense to – and we should – explore the alternative and potentially more explanatory solution explaining why feeling presently exists.

Since epiphenomenalism should only be a last resort, the assumption we have to work on is that feeling has some biological causal function, i.e., that feeling can contribute to a living being's overall biological fitness. Assuming that this is the case, we could make sense of why feeling has been preserved or even fostered by natural selection and, most importantly, of why it presently exists. In fact, it seems reasonable to suppose that feeling has been preserved throughout evolution by natural selection until today because it somehow contributed positively to the biological fitness of the ancestors of the presently feeling species. Once we assume *that* feeling has causal powers, however, the difficult question is explaining *how* feeling could influence biological fitness. This is the challenge I tackle in chapter 7.

7 A Hypothetical Biological Function of Feeling

In chapter 6 I have outlined the strategy I intend to use to suggest an explanatory solution to the Natural Problem, arguing there are reasons to suppose that feeling probably has a biological causal function (henceforth "biological function"). The present chapter is aimed at the difficult challenge of explaining *how* feeling could influence biological fitness. If this challenge is met, this could provide a viable explanatory answer to the Natural Problem. I begin by distinguishing two methodological strategies one might adopt to tackle the challenge (section 7.1). I consider the bottom-up approach but I argue that it is not adequately equipped to face the challenge (section 7.2). I thus adopt a top-down strategy elaborating in detail a hypothesis suggesting *how* feeling could influence biological fitness (section 7.3).

7.1 How To Settle the Question: A Methodological Point

In order to determine what the biological function of feeling might be (or have been) and how feeling could increase (or have increased) some beings' biological fitness, we could refer to different methodological strategies: bottom up and top down.

The bottom-up strategy adopted to determine the biological function of consciousness consists in comparing specific cases in which particular mental states are conscious with cases in which they are not conscious, trying to construct a theoretical account of the biological function of consciousness on the basis of such observations. By means of a contrastive method (see section 1.3) one can empirically observe what is missing in terms of a being's capacities when a mental state is not conscious, as opposed to when the same mental state is conscious. By analysing such differences, one can then construct a hypothesis suggesting what could be the biological function of that mental state being conscious.

Correlations between deficits in transitive consciousness (i.e., whether one is conscious *of* a given target) and deficits in functional capacities may be useful in constructing hypotheses about the biological function of those specific kinds of mental states being conscious. The bottom-up strategy has however two major limitations. *First*, a correlation does not necessarily imply a direct causal connection between the *relata*. That is, even though a given change in functional capacities might be shown to correlate with a given change in transitive consciousness, it does not follow that the cause of the change in functional capacities occurs by virtue of the change in consciousness. There might be alternative explanations for the correlation[143]. Thus, a bottom-up method alone is insufficient to settle the question of the function of specific kinds of consciousness. What is needed in addition to this is a theoretical account explaining why we could conclude that the correlations between changes in functional capacities and changes in consciousness are indeed due to a direct causal connection, rather than to something else.

Moreover, even if a convincing theoretical account in this direction was found, a *second* limitation of the bottom-up strategy is that it would be bound to draw conclusions limited to the specific correlations observed. That is, it might only explain what the biological function of – say – being visually conscious of a target stimulus is, but it is not justified to generalize this result and draw the conclusion that that is the biological function of consciousness *simpliciter* (of intransitive consciousness). Even if we allow that it is sound to talk about the biological function of particular mental states being conscious (rather than global states of mind being conscious) – something I am skeptical about (see chapter 4, and in particular section 4.1) – by operating bottom-up and focussing on the function of *transitive* consciousness, one cannot expect to find a direct answer to the question of what is the biological function of being conscious *tout court*. In order for such a generalization to be justified, one would have to provide extra evidence and a supportive argument explaining *why* the various particular biological functions that transitive consciousness is taken to have in every specific scenario could jointly show that being conscious in the intransitive sense has a given biological function. The two limitations just outlined are significant – probably even decisive – obstacles keeping the bottom-up methodological strategy from explaining what the biological function of feeling might be (or have been) and how feeling could increase (or have increased) some beings' biological fitness.

Instead of proceedings bottom-up, one could operate by means of a top-down strategy. That is, one could begin by directly proposing a theoretical hypothesis

---

143 A slightly different argument along these lines is the one developed by Block (1995, p. 242) on the basis of his distinction between phenomenal consciousness (P-consciousness) and access consciousness (A-consciousness). Block considers cases in which both consciousness (in its unspecified, ambiguous sense) and some functional capacities are missing, and holds that it is a fallacy to slide "from the premise that "consciousness" is missing – without being clear about what kind of consciousness is missing – to the conclusion that P-consciousness has a certain function". Block rightfully points out that the correlation might not be between P-consciousness and the missing functions in question, but rather between the latter and A-consciousness. My point differs in that I want to claim that any correlation between lack of consciousness and functional deficit (even if rightfully individuated) is not alone sufficient to conclude that the cause of the functional deficit is the lack of consciousness.

suggesting what the function of being conscious *simpliciter* might be, and – eventually – expose the hypothesis to testing in order to see whether it is sound and corroborated by empirical data. One way of outlining this strategy consists in suggesting that if feeling has a biological function at all, it has a *general* biological function (i.e., a biological function of being in a phenomenally conscious state of mind as opposed to being in an unconscious state of mind) and not a number of particular biological functions which are specific to particular kinds of mental states being conscious.

Both the bottom-up and the top-down strategy have a reason to exist and have their own limitations, but operating in a bottom-up way requires bowing to the theoretical assumption about the nature of the mind imposed by the contingent limitations of the scientific study of consciousness. The "contingent limitations" consist in the fact that salient empirical results obtained by means of a contrastive method, as already suggested, can only be *correlations* between functional differences and consciousness differences in specific instances of transitive mental states (e.g., conscious visual perception *of* something VS unconscious visual perception *of* something). Clearly, it is interesting to know what functional differences we can observe between cases in which one has a conscious visual perception *of* something, and cases in which one is not visually conscious *of* the same thing. However, as I have highlighted above, this limits the scope of the quest for the biological function of consciousness to very specific and individual kinds of *transitive* consciousness. Since going from there to a general hypothesis about the biological function of being conscious *simpliciter* is at the very least problematic, I briefly survey some suggestions of the bottom-up approach in section 7.2, but then I proceed by means of a top-down approach to the Natural Problem (section 7.3).

7.2 Bottom-up Approach

Fred Dretske (1997, p. 4) claims that questions about the function, the advantages, the purpose, the benefits, or the good of being conscious can be either questions about creature consciousness or about state consciousness. According to him the answer to the question about the function of *creature* consciousness is obvious:

> Let an animal – a gazelle, say – who is aware of prowling lions – where they are and what they are doing – compete with one who is not and the outcome is predictable. The one who is conscious will win hands down. Reproductive prospects, needless to say, are greatly enhanced by being able to see and smell predators. *That*, surely, is an evolutionary answer

> to questions about the benefits of creature consciousness. Take away perception – as you do, when you remove consciousness – and you are left with a vegetable. You are left with an eatee, not an eater. That is why the eaters of the world (most of them anyway) are conscious. (Dretske 1997, p. 5, original emphasis)

This, for Dretske, is sufficient to show that creature consciousness is good for something (i.e., that it has a biological function). Dretske's analysis, however, might come too quickly and be deceiving. First notice that he considers an example of *transitive* creature consciousness – the gazelle being aware *of* prowling lions – concluding that having a conscious perception *of* the lions is better than not having a conscious perception *of* the lions. One could grant him that and still resist the conclusion that what makes a difference is the fact that the gazelle is *phenomenally conscious* of the lion. More precisely, one could contend that what Dretske mistakenly takes to be a "benefit" of transitive being consciousness – namely, making the gazelle aware *of* the lion – is in fact a benefit of having a given brain state that happens to correlate with a given global *quale*. If this is the case, then the fact that the gazelle is in a transitive phenomenally conscious state of mind (i.e. able to *consciously* see and smell predators) is not strictly speaking an explanation for the enhanced reproductive prospects of the gazelle's; it just happens to be contingently correlated to what really, at a more fundamental level, enhances the reproductive prospects of the gazelle. In short, it is not clear why the fact that the gazelle is *phenomenally conscious* of the lion has to be an explanation of the gazelle' success since what makes a causal difference may very well be the subserving base on which the conscious state supervenes (which is absent in the case in which the gazelle is not conscious of the lion).

Moreover, endorsing Block's terminology, we might cast the above example as a case in which the term "consciousness" is adopted in the unspecified sense conflating access consciousness with phenomenal consciousness, and where an apparent function of access consciousness is mistakenly presented as a function of phenomenal consciousness. According to Block (1995, p. 231) a state is A-conscious if, by virtue of one's having the state, "a representation of its content is (1) inferentially promiscuous (Stich 1978), that is, poised for use as a premise in reasoning, (2) poised for rational control of action, and (3) poised for rational control of speech". If this is correct, it seems that all the lion needs in order to be successful is to be A-conscious, and that the functional correlations to which Dretske is pointing could be correlations between A-consciousness and given functional capacities of the lion (e.g., see, smell, hunt, and so on). It is unclear why differences in the fact of being in a phenomenally conscious state of mind should be taken to be the *cause* of any functional difference, given that all we can observe is that they correlate with changes in biological functionality. This, however, is

Dretske's suggestion, since he explicitly says that the function of *experience* (i.e., phenomenal consciousness) is to enable conscious beings to do all those things that non-conscious beings cannot do:

> The function of experience, the reason animals are conscious of objects and their properties, is to enable them to do all those things that those who do not have it cannot do. This is a great deal indeed. If we assume [...] that there are many things people with experience can do that they cannot do without it, then that is a perfectly good answer to a question about what the function of experience is. That is why we, and a great many other animals, are conscious of things and, thus, why, on an act conception of state consciousness, we have conscious experiences. (Dretske 1997, p. 14)

Again, the problem is that it is not clear that what enables conscious animals to do all those things that non conscious animals cannot do really is *phenomenal* consciousness. Dretske, interestingly, points out that if something else besides experience could enable us to do the same things, this would only show that there is more than one way to get the job done, but not that experience does not have a function (1997, p. 14). However, I claim, this is not true. If phenomenal consciousness is epiphenomenal, for example, it would not have the biological function that Dretske assumes it has. In order to grant Dretske's point, one would first have to give reasons suggesting that feeling has a biological function, and I think that Dretske's argument falls short in this sense. We can make this clear by adapting an observation by Block (1995, p. 240) originally arising as criticism of Searle's suggestion about the function of consciousness: for any argument for the function of P-consciousness to have any force, "a case would have to be made that P-consciousness is actually missing, or at least diminished" when given functions are missing. If I am right in saying that Dretske uses the mongrel concept "consciousness" conflating P-consciousness and A-consciousness, then Dretske's claim that creature consciousness obviously has a function is ambiguous. Dretske seems to interpret the examples as showing that creature consciousness intended as *P-consciousness* (i.e., feeling) has a function. If my interpretation is correct, however, Dretske's inference to the latter conclusion is fallacious, and the answer to the question about the function of creature consciousness, intended as P-consciousness, is not as straightforward as he takes it to be. In the example presented by Dretske, in fact, what is apparently missing when given functions are missing seems to be access to some information (i.e., whether there is a lion). This might provide support for the claim that creature consciousness intended as A-consciousness has a function – but not for the claim that feeling has a function. Since answering the Natural Problem requires telling how feeling (i.e., being *phenomenally* conscious) could influence fitness, Dretske's analysis seems to be unsatisfactory.

Unaware of the potential problem just mentioned, Dretske claims that the question about the biological function of states, processes, and activities being conscious is really problematic only on the backdrop of higher-order theories endorsing an object conception – rather than the first-order act conception – of state consciousness. According to higher-order theories, in fact, beings can be conscious of things without ever occupying a conscious state (e.g., one can be conscious of a lion-experience even without having conscious lion experiences):

> If what makes an experience or a thought conscious is the fact that S (the person in whom it occurs) is, somehow, aware of it, then it is clear that the causal powers of the thought or experience (as opposed to the causal powers of the thinker or experiencer) are unaffected by its being conscious. Mental states and processes would be no less effective in doing their job [...] if they were all unconscious. (Dretske 1997, p. 8)

If conscious mental states do not differ in some causally relevant way from unconscious mental states, then, as Dretske (1997, p. 2) says, "no good comes from being conscious", and it is difficult to understand why conscious states presently exist and how they could have a biological function. If the fact that a mental state is conscious does not make a difference to what the mental state does – if a mental state is not enhanced (or diminished) by the fact of being conscious –, then nothing a mental state does can be explained by its being conscious, and – as Davies and Humphreys (1993, pp. 4–5) say, "psychological theory need not be concerned with this topic" (i.e., with the biological function of conscious mental states). I think that Dretske's analysis on this point is correct, but it should be highlighted that, even by adopting a first-order act conception of state consciousness, we might end up misestimating the problem of knowing whether feeling has a biological function. This, however, is not to say that I have yet put forward a satisfactory alternative hypothesis about the biological function of feeling that could explain why feeling presently exists.

Something that I find unsatisfactory in Dretske's characterization – besides the points already mentioned – is that he considers consciousness only in its transitive form (i.e., being conscious *of* something) – saying that the biological function of consciousness is to enable beings who are conscious *of* things to do all those things that beings who are not conscious *of* things cannot do. This implies a "state consciousness" approach suggesting that there could be many different biological functions of consciousness relative to different particular kinds of conscious experiences *of* something one might have. For example, there might be a particular biological function of being visually conscious *of* something, a particular biological function of being auditorily conscious *of* something, and so on. The biological function of *visual* consciousness would be to enable beings that are *visually* conscious *of* things to do all those things that beings that are not *visually* conscious *of* things cannot do, and the same thing for the other sensory modalities. As I have already stated, however, I believe that individuating particular functions of transitive kinds of consciousness of this sort is not alone sufficient to give a satisfactory reply to the question of knowing what is the biological function of consciousness *simpliciter* (i.e., of being phenomenally conscious) – and this is what is required in order to answer the Natural Problem.

Let us consider an example of how a bottom-up approach tackles the question of identifying possible biological functions of consciousness. Dretske (1997, p. 12 ff), in order to suggest what a function of consciousness could be, considers the special case of blindsight. He questions what a blind-sighted being is not able to do if compared to a being with normal conscious vision *of* its surroundings. On the basis of the information gathered by the scientific study of the specific case of blindsight, Dretske deduces what a possible biological function of conscious vision (as opposed to blindsight-like vision) may be.

Blindsight[144] is a condition in which subjects have blind areas in their visual field (usually, in one side of their visual field) as a result of cortical lesions in the primary visual cortex (also known as area V1 or striate cortex). If we flash a stimulus in a blind area of blindsighted subjects and ask them what they saw, they reply that they did not consciously see anything. This is usually interpreted as suggesting that in those cases consciousness is missing[145]. Surprisingly, however, if asked to guess, some (but not all) blindsighted subjects are nonetheless able to reliably discriminate some features of the stimuli connected to location, direction and motion. That is, they are able to obtain some information about their nearby spatial environment (e.g., that there is an X and not an O on the left), without thereby having the corresponding conscious visual experience (e.g., of no *qualia* of the X on the left). For example, if some blindsighted subjects are asked to grasp an object in their blind field, even though they cannot consciously see it, they can shape their hands in a way appropriate to grasp the object in question. Furthermore, blindsighters are able to walk around avoiding obstacles despite not consciously seeing them. In short, there is abundant evidence that blindsighters can obtain some information required to determine appropriate action without being conscious *of* (most) properties of the space surrounding them. What is usually taken to be missing in blindsight, however, is the ability to deploy such information in reasoning, reporting, and rational control and guidance of action.

---

144 Blindsight was originally presented by Pöppel et al. (1973) and it has been increasingly discussed in the consciousness literature ever since. For more on this phenomenon see Campion et al. (1983); Weiskrantz (1986, 1997); Milner and Rugg (1992); and Bornstein and Pittman (1992).
145 For an opposite view, see McGinn (1991, p. 112).

The "rational" clause, as Block (1995, p. 228) suggests, is meant to exclude the "guessing" kind of guidance of action that blindsight patients are capable of. For example, Marcel (1986) claims that thirsty blindsight subjects would not spontaneously reach for a glass of water in their blind field. This is taken to suggest that the information unconsciously represented in the blindsighted subject's brain is not deployed rationally to guide the action of drinking. A common way to interpret this, as Block puts it, is that "when a content is not conscious – as in the blindsight patient's blind field perceptual contents, it can influence behavior in various ways, but only when the content is conscious does it play a *rational* role; and so consciousness must be involved in promoting this rational role" (1995, p. 228, original emphasis). In other words, since in blindsight the information represented in the brain is not used in reasoning, reporting, and rationally controlling and guiding action, and since in blindsight consciousness is missing, it is tempting to claim that the function of consciousness consists precisely in enabling the information represented in the brain to be used in reasoning, reporting, and rationally controlling and guiding action[146].

Dretske (1997) considers the case of the blindsighted monkey Helen. Helen developed a capacity to move through a room full of obstacles – supposedly – conceptually representing where objects were, despite not having conscious visual perceptions of them being there. Here is the original characterization by Humphrey (1970; 1972; 1974; 1992):

> I do want to suggest that, even if [Helen] did come to realize that she could use her eyes to obtain visual information, she no longer knew how that information came to her: if there was a currant before her eyes she would find that she knew its position but, lacking visual sensation, she no longer *saw* it as being there. [...] The information she obtained through her eyes was "pure perceptual knowledge" for which she was aware of no substantiating evidence in the form of visual sensation. (Humphrey 1992, p. 89, original emphasis)

Dretske, generalizing from this case, suggests what the function of being conscious *of* something might consist in. He suggests that consciously experiencing (i.e., seeing, hearing, smelling) objects might help our identification and recognition of them:

> Remove visual sensations of X and S might still be able to tell *where* X is, but S will not be able to tell *what* X is. Helen couldn't. That is – or may be – a reasonable empirical conjecture

---

[146] For such positions see Baars (1988); Flanagan (1991; 1992); Marcel (1986; 1988); Van Gulick (1989). For a different bottom-up view holding that *qualia* facilitate non-automatic, decision-based action see Ramachandran and Hirnstein (1998).

for the purpose of experience[147] – for why animals (including humans) are, via perceptual experience, made aware of objects. (Dretske 1997, p. 13, original emphasis).

Dretske, in sum, claims that there are many important facts we cannot know unless we are consciously aware of objects these facts are facts about (1997, p. 12)[148]. Dretske's observation that being visually conscious *of* a target (rather than not being visually conscious *of* it) seems to play a role in object identification and recognition might indeed interestingly indicate a possible function of having transitive visually conscious mental states. However, we are not justified in drawing general conclusions about the biological function of feeling *simpliciter* only by independently considering very specific and restricted cases such as the possible biological function of having transitively visually conscious mental states.

Another suggestion coming from bottom-up approaches is that consciousness has the function of promoting flexibility and creativity[149]. This conclusion comes from the analysis of cases of petit mal epileptics that have seizures while driving, walking or doing other automatized activities such as playing the piano. The epileptic subjects in questions are described as continuing their activities in a mechanical and routinized manner during seizures despite being "totally unconscious" or converted into "mindless" automata (cf. Penfield 1975, pp. 37–40). Since during the seizures these subjects seem to lack both consciousness and behavioural flexibility and creativity, it is suggested that this correlation indicates a possible function of consciousness. Searle says:

> The epileptic seizure rendered the patient totally unconscious, yet the patient continued to exhibit what would normally be called goal-directed behavior. [...] Now why could all behavior not be like that? Notice that in the cases, the patients were performing types of actions that were habitual, routine and memorized [...] normal, human conscious behavior has a degree of flexibility and creativity that is absent from the Penfield cases of the unconscious driver and the unconscious pianist. Consciousness adds powers of discrimination and flexibility even to memorized routine activities. [...] One of the evolutionary advantages conferred on us by consciousness is the much greater flexibility, sensitivity, and creativity we derive from being conscious. (Searle 1992, pp. 108–109)

---

[147] Notice the reference to the "purpose" of experience. This sort of terminology is an undesirable reminiscence of a teleological way of thinking about biological functions (see my previous comment on this in section 3.3).
[148] A particularly interesting consequence of this, as I already said, is that if we are not consciously aware of something we are not able to initiate intentional action with respect to it (Marcel and Bisiach 1988). Blindsight patients can guess the presence or absence of some objects and interact appropriately with them in a forced-choice situation, but they would not spontaneously initiate any such action in normal life situations.
[149] For such arguments see Searle (1992) and Van Gulick (1989, p. 220).

Again, the first problem with this conclusion is that – as Block (1995) rightly points out – Searle is giving no reason to believe that *P-consciousness* is missing or even diminished in intensity in the epileptic cases he considers. The behavioural inflexibility and lack of creativity observed might be explained solely by a deficiency in thought processes and A-consciousness. Block rightfully concludes that the ordinary mongrel notion of consciousness wrapping P-consciousness and A-consciousness together is being used, and that "an obvious function of A-consciousness is illicitly transferred to P-consciousness" (1995, p. 240). Furthermore, even if there were truly changes in *P-consciousness* that correlated with chances in flexibility and creativity in petit mal epileptic seizures, knowing this would not be alone sufficient to claim that being conscious *simpliciter* has the biological function of promoting flexibility and creativity.

Bottom-up empirically based approaches such as the ones sketched above are useful to point out correlations between particular deficiencies of transitively conscious mental states in specific modalities and functional deficiencies. By means of a contrastive method, bottom-up approaches allow the advancement of interesting hypotheses suggesting what the biological function of specific sorts of mental states being transitively conscious might be. However, there are several difficulties involved. First, if the target of the functional explanation is P-consciousness, one has to clearly show that a given functional deficit correlates to a case in which P-consciousness (and not just A-consciousness) is missing or is diminished[150]. Second, even if one can avoid this obstacle, correlations alone do not suffice to conclude that it is the fact of being P-conscious that makes a functional difference. Third, the contrastive method that – for perfectly understandable contingent reasons – the science of consciousness has to adopt for the functional study of transitively conscious mental states, is not alone sufficient to draw conclusions about the function of feeling simpliciter. That is, results from the contrastive functional analysis of conscious and unconscious mental states in specific sense modalities cannot be directly generalized to provide a conclusive answer about the biological function of being conscious *simpliciter*. Since bottom-up approaches are faced with several important obstacles, I section 7.3 I tackle the challenge of explaining how feeling could influence biological fitness by switching strategy and proceeding top-down by formulating a theoretical hypothesis about a possible biological function of feeling *simpliciter*.

---

150 For an impressively long list of attempts to identify the function of consciousness that apparently fail to do so see Block (1995, esp. pp. 236–237).

7.3 Biological Function: a Top-Down Hypothesis

A brief recap might be useful at this point. I am assuming that feeling has a biological function – i.e., that feeling can influence biological fitness. If feeling does positively contribute to a being's biological fitness it provides a selective advantage to feeling beings and this, I claim, would offer an explanatory answer to the question of knowing why some beings presently feel (i.e. the Natural Problem). So far so good. However, in order to justify such an answer to the Natural Problem I still need to explain (i) *what* the causal role of being conscious *simpliciter* might be, and (ii) why it is played by feeling (i.e., P-consciousness) rather than by something else. I have claimed (section 7.2) that the better-suited strategy for tackling these questions is a top-down strategy, thus this is how I proceed to develop my hypothesis.

How could feeling increase (or have increased) the biological fitness of a being? How could an ontologically subjective biological phenomenon make a causal difference that is not equally explainable in terms of its subserving ontologically objective neurophysiological base alone? The intuition I want to push is that in order to increase a being's biological fitness, feeling has to be closely connected with the strongest and most primitive forces behind evolution, that is, survival and reproduction. But, again this raises questions: How could feeling enhance one's chances of survival and reproduction? How could feeling drive one's behaviour in such a way that a feeling being is better equipped to avoid death and pursue reproduction than a non-feeling being? Last, but not least, why is the ontologically subjective dimension of feeling important? I approach these questions from afar, beginning by explaining what biological efficacy consists in and how living beings can – in general – achieve such efficacy (section 7.3.1).

7.3.1 Biological Efficacy and Strategies to Maximise Profit

Biological efficacy consists in obtaining on average a positive outcome (a benefit) in the compromise between energy invested in performing a task (i.e., its metabolic cost) and energy gained from it. To use a rather silly example, if on average a tiger invests more energy for hunting than it gains from it (since – say – it almost always fails to capture a prey and feed herself), that tiger exhibits a behaviour that is not biologically efficient and that lowers her biological fitness value. In the actual natural world, under the energy and time constraints imposed by the natural environment, any living being has to be able to react efficiently (quickly and well enough) to as many possible token configurations of input stimuli as possible. Only by maximizing one's energy outcome in most possible conditions can one ensure itself a high biological fitness (i.e., increase its chances of surviving at least until the production of fertile offspring).

There are two sorts of strategies that may in principle be enacted to maximise the energy profit. The first is an *algorithmic strategy* consisting in following a set of well-defined and exhaustive encoded instructions for carrying out any specific task and achieving an optimal result (*à la* Turing machine). The second is a *heuristic strategy* consisting in following a set of accessible and efficient (albeit not exhaustive) instructions for quickly carrying out any task and achieving a satisfactory (even if sub-optimal) result. Beings constantly faced with simple and stable being-environment interactions might benefit from a rigid fine-grained token input detection system triggering a standard fitness enhancing output response to a given input token. On the other hand, beings constantly faced with complex and varying being-environment interactions[151] might benefit from a "rough" type input recognition system detecting patterns of similarities between input tokens and *quickly* producing *good enough* responses. In fact, the more variable the being-environment interactions are, the more the amount of information that could have to be handled increases. Under the strong constraints of life in natural settings (where time and resources are not infinite) quick sub-optimal reactions to novel input token situations are on average more efficient than optimal reactions arriving too late.

We know from the science of consciousness that the input detection and the neural information processing work occur mostly unconsciously (cf. Dehaene 2014). However, it is a fact that we sometimes are conscious. I assume that when heuristic neural information processing occurs in beings with complex integrated multi-modal sensory systems, the information about a given token state of affairs is automatically and involuntarily subjectively represented onto a weighted, unified "rough" feeling token – a global quale. How that happens is the Hard Problem and does not concern me here. I just assume that this translation (or something of the sort) *does* happen and proceed to explain how assuming this could help in solving the Natural Problem.

---

[151] What distinguishes "simple" and "complex" being-environment interactions is not the mere quantity or variety of sensory modalities by which a being can access the environment (e.g., it does not matter whether one has only vision, or vision, chemoreceptors, and hair-sensitivity), but rather whether diverse information is or is not integrated. I thank Aaron Ancell for pointing this out during a talk at Duke University in February 2014.

7.3.2 Global *Qualia* and Multiple Realization

I suppose that in the actual natural world, for any individual being, at any given moment one has access to a finite set of information that is inferior to the global set of information about the world[152]. The information one *can* access depends on one's physical location and functioning of the different senses, environmental and contextual conditions, internal states, and other conditions of the sort. What matters is that every one of us has an access limited to a small portion (i.e., a subset) of the overall information that is potentially accessible at any one time. I call the set of information that an individual being accesses at any one moment "*token input stimulus*". Accordingly, I name different token input stimuli "tIS_1, tIS_2, ..., tIS_n".

I further suppose that at any given moment the token input stimulus that an individual being is accessing correlates with a state of the brain (that is, of course, in animals with a brain). I call such a state of the brain "*token global neural state*". A token global neural state can be pictured as a specific temporal section – a snapshot, if you prefer – of the overall (global) brain activity of a being at a given moment. The correlation between the token input stimulus and the token global neural state holds because one is in that specific token global neural state *by virtue* of accessing that specific token input stimulus – i.e., that specific set of information – rather than another one. I name different token global neural states "$n_1, n_2, ... n_n$".

Finally, I suppose that some, but not all, token global neural states are the supervenience base for phenomenally conscious states of mind. That is – in accordance to supervenience and the Mind-Body dependence thesis – I suppose that at any given moment, one's token global neural state fully determines whether one is or is not conscious *simpliciter*, i.e., whether one has or does not feel. I name different feeling tokens "$f_1, f_2, ..., f_n$".

To sum up so far, for any individual being at any one time the token input stimulus (i.e., the set of accessed information) determines one's token global neural state, that in turn – if appropriate – may be the supervenience base for a feeling at that time. Let us sketch two scenarios as an example (Table 8).

In scenario (1), accessing the token input stimulus tIS_1 determines a being's token global neural state n_1 which is not a supervenience base for any feeling token. In this scenario the being in question is not conscious *simpliciter*, i.e., it is in an unconscious state of mind. In scenario (2), accessing the token input stimulus tIS_2 determines a being's token global neural state n_2, which – let us

---

[152] That is, in the actual natural world there are no omniscient God-like beings.

suppose – is a supervenience base for feeling f_2. In this scenario, contrary to the former case, the being in question *is* conscious *simpliciter*, i.e., he is in a phenomenally conscious state of mind. Let me stress again that explaining *why* feeling occurs in addition to (or supervenes on) a given token global neural state is not my concern here – I am just assuming this to be the case.

Table 8

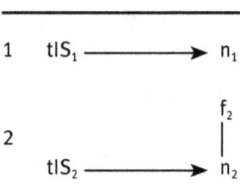

A further assumption that I have already previously outlined is that any feeling that a being can have has necessarily a global *quale* – indeed it *essentially consists* in its global *quale*. In other words, for any being in a phenomenally conscious state of mind, there is necessarily something it (globally) feels like to be in that state of mind, something it feels like *for* the being. I believe that if feeling has a biological function at all, then such a function has to depend on the ontologically subjective qualitative dimension that characterizes this biological phenomenon. For this reason, I consider feelings more in detail.

Any occurrence of feeling has a specific, distinct, and unique *global quale*. The way in which the feeling feels like, as I already said, is determined by the subserving token global neural state (i.e., the ontologically objective physical/functional state that acts as supervenience base for the feeling) which in turn is reliably determined by the accessed token input stimulus. Thus, even if indirectly – because of transitivity –, any feeling feels the way it does (rather than in another way) as a consequence of being caused by a token input stimulus rather than by another.

In accordance with the supervenience claim (cf. section 2.2.3) I hold that identity at the neural level implies identity at the feeling level. That is, if two beings are exactly in the same token global neural state they also have the same feeling token (i.e., the same global *quale*)[153]. This implies that there can be no

---

[153] When I say "identity at the neural level" I am not committing specifically to either physical identity or functional identity. The identity that matters is the one appropriate to the task – whichever it is.

variation in feeling without a corresponding variation at the neural level. Let us sketch two scenarios as an example (Table 9).

Table 9

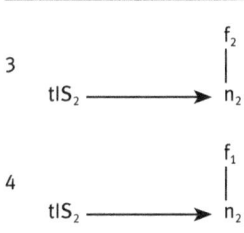

If scenario (3) is true – i.e., if feeling token f_2 supervenes on token global neural state n_2 – then scenario (4) cannot be true (and vice versa), since there cannot be a change in feeling (from f_2 to f_1) without a corresponding change of the supervenience base.

On the other hand, I claim that identity at the feeling level does not necessitate an identity at the neural level. That is, if two beings have the same feeling token (i.e., the same global *quale*) they may or may not be in the same token global neural state. This comes down to claiming that any specific feeling can – at least in principle – be multiply realized by different token global neural states. A feeling token f_1 (with a specific *global quale*) can be multiply realized by a number of objectively differing token global neural states $n_{1,2,3,n}$. This comes down to also saying that – because of transitivity – a feeling token f_1 can occur in response to objectively differing token input stimuli. The idea is that we can remain *phenomenally* blind to objective variations – usually, *minor* variations – of token input stimuli even if we could in principle detect such variations at the neural level[154].

Consider some scenarios as example (Table 10).

According to the above hypothesis, it is possible for same feeling token f_1 to be multiply realized by different token global neural states n_1, n_2 and n_3. Thus

---

[154] This is a metaphysical claim, and it is not up to me to claim whether this is or is not *actually* the case (i.e., how the actual world is like with respect to multiple realization). The problem, already mentioned, is that there is an epistemological limitation to what we can know about other being's feelings. Notice that the multiple realization claim leaves open the possibility that different individual beings of the same or different species could in principle multiply realize exactly the same feeling token if in the appropriate token global neural state.

scenarios (5), (6) and (7) can all be true. There is a contingent limitation to which global neural state tokens can realize a single feeling token, but knowing what that limitation could be is not required for our purposes. What matters is simply admitting that there could be changes at the neural level that do not result in changes at the phenomenal level, and that a single feeling token can be multiply realized by objectively differing global neural state tokens.

Table 10

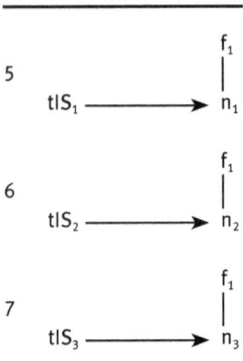

I claim that the reason why multiple realization may occur is that feelings (i.e., global *qualia*) are weighted, unified and "rough". That is, feelings are ontologically subjective abstractions that do not phenomenally represent the ontologically objective token input stimulus conveyed by the token global neural state in all of its detail and fineness of grain; this despite the fact that feelings might *seem* to be a faithful and transparent phenomenal representation of the world. Feelings, I claim, are a succinct phenomenal synthesis of the salient information that the being may require for action. Thus, if there is no difference in *salient information* between two scenarios, despite there being *de facto* objective differences, the latter may go entirely unnoticed at the phenomenal level.

An example might be useful to illustrate this point. Imagine a person, Amanda, who is sitting in a silent white room and looking at a Luis Royo painting. At time t_1, Amanda is objectively detecting a token input stimulus tIS_1, the totality of the information that she can access in that moment (i.e., all the information she has access to by means of vision, audition, touch, odour, taste, proprioception, memory, and so on). Amanda has a neural state n_1 that is normally connected to the token input stimulus tIS_1 and – let us suppose – a given feeling f_1 that supervenes on n_1. Now we slightly change the objective token input stimulus by adding a black dot to the painting Amanda is looking at. Let us admit that everything else remains unaltered. In this second scenario Amanda will be confronted to a different token input stimulus tIS_2, will have an objectively different neural state n_2 (since the input information has objectively changed), but might nonetheless have exactly the same feeling f_1 she previously had. The reason for this, I claim, is that the token input situation has undergone an objective change, but has not changed drastically nor in a salient manner and, accordingly, the token global neural state has undergone only a minor change that in this case might not be sufficient to alter the feeling supervening on it. Let us now imagine a third scenario in which we drastically change the token input stimulus. We paint the room in black, turn on a stereo playing "Bohemian Rhapsody" by *Queen*, and substitute the Royo painting with the statue *Three Man Walking II* by Alberto Giacometti. Now Amanda is confronted to a *radically* objectively different token input stimulus tIS_3. As a consequence, I claim, she will have a neural state n_3 drastically differing from both n_1 and n_2 (since the input information has changed in important respects). This drastic difference, I claim, is salient. It would therefore be reasonable to suppose that Amanda in this situation could have a feeling f_3 that differs *phenomenologically* from the previous feeling f_1 – i.e., a feeling that has a different global *quale*. This of course is just a thought experiment, but the point is to show that, if this is plausible, we can make sense of the idea that minor changes at the ontologically objective level might go completely unnoticed at the ontologically subjective level. In normal circumstances minor non-salient objective changes in the detected token input stimulus do not entail a phenomenal change, whereas drastic salient changes in the token input stimulus are mapped as phenomenal changes at the ontologically subjective level[155]. This is all good, but how is multiple realization of any help?

7.3.3 Similarities Across Global *Qualia*

An interesting and unique feature of conscious experience is that even if two global *qualia* are phenomenologically, subjectively, different – i.e., even if two feeling tokens are qualitatively distinct – they can subjectively appear to share some (albeit not all) qualitative properties[156]. For example, even though the

---

[155] The phenomenon of change blindness supports the hypothesis that an objective change in the input can go unnoticed at the conscious level, even though change blindness is presented in the context of transitive "state consciousness" (i.e., as not being visually conscious *of* an objective change in the visual field).

[156] This idea could be developed to suggest a mereological theory of conscious experience ac-

global *qualia* of toothache experiences are phenomenologically different from the global *qualia* of headaches (e.g., since the perceived quality and phenomenal location of the painfulness differs), toothache and headache feelings share the qualitative property of being painful (or disturbing, annoying, and so on). Interestingly, phenomenologically, we would describe a toothache feeling as "more similar" to a headache feeling than to a feeling of excitation or to a feeling of sheer joy by virtue of the fact that the former two share a similar qualitative property (i.e., painfulness). The presence of the "painful" qualitative property across different toothache and headache feeling tokens is constant in time and stable across subjects. That is, it is not just random: in normal conditions both toothache and headache feelings are always painful. This suggests that even though different global *qualia* (such as toothache and headache feelings) are phenomenologically distinct by virtue of their differences in global phenomenal character, they might nonetheless share *some* (but not all) qualitative properties. On the basis of this observation, I suggest that different feeling tokens might reliably be regrouped in feeling types (e.g., "painful feelings") on the basis of specific qualitative properties that they have in common (i.e., on the basis of specific subjectively salient resemblance criteria). I name different feeling types "$F_1, F_2, ..., F_n$".

A somewhat similar note applies also to token global neural states. Even if two objectively different token global neural states are objectively different, they can share some or even most (but not all) of their physical and/or functional properties, and therefore they might be regrouped as belonging to the same type of global neural states. I name different types of global neural states "$N_1, N_2, ... N_n$".

Interestingly, the way in which *feeling* types are individuated differs from the way in which *global neural state* types are individuated. The former can only be individuated subjectively, by means of *phenomenally salient resemblances* among different feeling tokens, whereas the latter can only be individuated objectively, by means of *objectively detectable physical and/or functional similarities* among different global neural state tokens. If there is a correlation between feeling types and global neural state types the above would be just an epistemological remark about how we know and/or individuate ontologically subjective types as opposed to ontologically objective types. If however the types of ontologically subjective feelings that we can phenomenally identify do not have an isomorphic counterpart in terms of ontologically objective types – that is, if there is no type-type identity – then the above might lead to an interesting remark about ontology. If

---

cording to which the proper parts of global qualia are those phenomenal properties that are recurrent across different global qualia, but are never found in isolation. I would call these phenomenal parts "disintegrated *qualia*".

there is a mismatch or asymmetry between feeling types and global neural state types then some feeling tokens might be ontologically related to each other not in virtue of underlying similarities between their respective supervenience bases, but rather solely by virtue of phenomenally salient resemblances among them.

My hypothesis about the biological function of feeling relies precisely on the above possibility. More precisely, it relies on the idea that different feeling tokens might share some of their qualitative properties despite the absence of any parallel similarity among the global neural state tokens on which they individually supervene. After all, there is no in principle reason to suppose that a subjectively perceived qualitative similarity between any two feelings necessarily requires an ontologically objective physiological or functional similarity between their respective subserving neural states; two feeling tokens could share some of their qualitative properties even if they have radically different neural activity subserving them. Importantly, I am not claiming that there *actually is* a mismatch between feeling types and neural types. I simply claim that, insofar as there is no conclusive reason to think that this is *not* possible, it is reasonable to construct a hypothesis conditionally on this possibility. As far as I can, tell there is no apparent contradiction in claiming this. Whether there *actually* is a mismatch between feeling types and neural types is open to debate, but as of now there is no conclusive argument against the possibility of such a mismatch, and this logical space is all I need.

7.3.4 The Example of Valence

For the sake of my hypothesis, I focus exclusively on the analysis of a single qualitative property – valence. I consider the notion of "valence" as applying to global *qualia* (i.e., to feelings). This is not the usual way in which such a term is used in the valence literature, since valence is usually discussed in the context of emotions[157]. Referring to a useful distinction by Charland (2005a; 2005b), I am clearly not interested in the positive or negative character of an emotion *simpliciter* (i.e., in the "emotion valence" of fear, joy, and so on). My use of "valence" is arguably

---

[157] For an excellent comprehensive introduction and overview of different conceptions of valence see Colombetti (2005). For defences of the thesis according to which emotions – at least basic emotions – are feelings of bodily changes see Damasio (1994); James (1884); Prinz (2004a, 2004b). Interesting contributions on the relation between feelings of emotions appear in an issue of the *Journal of Consciousness Studies* entirely dedicated to emotion experience (Prinz 2005; Ratcliffe 2005; Russell 2005). I thank Rebekka Hufendiek for suggesting me this literature.

more in the direction of "affect valence" in emotion theory, that is, the positive or negative character of an *emotion experience* such as experiencing fear, joy, and so on. However, I consider "valence" as a qualitative property of phenomenally conscious states of mind in general, rather than as the positive or negative character of emotion experiences only. Perhaps valence feelings as I intend them could be considered emotions, but – also considering that the notion of "feeling" is not univocal – this is a controversial claim. Given that answering the Natural Problem does not require taking a stance in the complex debate on the relation between feelings and emotions, I do not commit to any such view and use valence as defined in what follows.

I hold that valence is a qualitative property that different feeling tokens (i.e., global *qualia*) can share and that might not be traceable or reducible to objective similarities at the level of the global neural state tokens underlying such feelings. Whether valence is really not traceable at the neural level is a crucial question, but it is an *empirical* question.[158] The hypothesis begins by suggesting that some – but not all – weighted, unified "rough" feeling tokens (i.e., some global *qualia*) have either a positive or a negative valence. Any feeling token that "feels good" can be described as having a *positive valence* (or an intrinsic attractiveness). On the other hand, any feeling token that "feels bad" can be described as having a *negative valence* (or an intrinsic "aversiveness"). Joyful, pleasurable, and satisfactory feelings are paradigmatic examples of feeling tokens that share the qualitative property of positive valence. On the other hand, fearful, disgusting and painful feelings[159], and so on are paradigmatic examples of feeling tokens

---

158 A recent extended meta-analytic study of neuroimaging literature on the brain basis of positive and negative affect (Lindquist et al., 2016) has tested three competing hypotheses concerning the brain basis of emotional affect valence. Little evidence was found for the *bipolarity hypothesis* (cf. Wundt 1897; Barrett and Russell 1999; Carroll et al. 1999; Larsen and Diener 1992) suggesting those positive and negative affects are supported by a brain system that monotonically increases and/or decreases along the valence dimension. Similarly, little evidence was found for the *bivalent hypothesis* (cf. Cacioppo et al. 1999; Larsen et al. 2001, 2004; Norris et al. 2010; Watson and Tellegen 1985) suggesting that positive and negative affects are supported by independent brain systems. The authors of the study conclude that the findings supported the *affective workspace hypothesis* (Barrett and Bliss-Moreau 2009) suggesting that positive and negative affects are supported by a flexible set of valence-general limbic and paralimbic brain regions. This hypothesis is worth investigating further, but for the time being the empirical evidence is insufficient to draw a convincing conclusion on this matter. I thank Rebekka Hufendiek for indicating me this literature.

159 In an influential article (1978c), Dennett suggested that there are clinical pain syndromes – what he calls "reactive dissociation" cases – in which pain affect (i.e., the valence of a pain experience) and the sensory aspect of pain are separated. This may suggest that pain experi-

that share the qualitative property of negative valence. I suggest classifying all the global *qualia* that do not have either a positive or a negative valence as having a *neutral valence*: they feel neither good nor bad. Valence should not be seen as an all-or-nothing qualitative property, but rather as something coming in degrees. We can try to conceive a "valence spectrum" ranging from positive valence, via neutral, to negative valence, on which any feeling token (i.e., any global *quale*) could be in principle mapped (Figure 3).

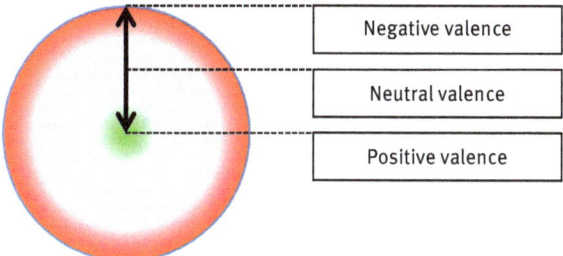

Figure 3

In this illustration the epicentre represented in green signals positive valence. The positive valence can come in progressively decreasing degrees, merging with neutral (or no) valence represented by the white inner ring. The external border of the circle represents negative valence, which can also come in progressively decreasing degrees, until merging with neutral valence[160]. The reason why I represent the valence spectrum on a bi-dimensional circular surface – call it "global *qualia* space" – is to suggest that any two feeling tokens (i.e., global *qualia*) that share the same valence – despite not sharing *all* their phenomenal properties and therefore being mapped at different locations in the global *qualia* space – can be mapped as similarly equidistant from the centre. This indicates that a similar valence can be shared by many different global *qualia*, regardless of other differences characterizing and distinguishing them at the phenomenal level.

---

ences do not *essentially* have a negative valence (i.e., they are not *essentially* painful, unpleasant and awful), but is perfectly compatible with the claim that, in normal non-pathological cases, pains do have a negative valence. My claim here is simply that in *normal* cases, pain, disgust, fear, and so on *do* have a negative valence. For a thorough discussion of pain, including a discussion of Dennett's argument, see Aydede (2013).
160 The reason I put positive valence in the centre and negative valence on the boarder, rather than *vice versa* is to suggest that positive valence signals homeostatic balance, whereas negative valence signals homeostatic imbalance.

I suggest this in Figure 4 by mapping different feeling tokens (represented by the star symbols), on different locations of the global qualia space and by showing that three distinct couples of feeling tokens can be individuated on the basis of valence similarity. Every couple of feeling tokens that is similarly equidistant from the epicentre of the global qualia space shares the same valence (i.e., the two green stars share a positive valence, the two white stars share a neutral valence, and the two red stars share a negative valence) despite having different global *qualia*.[161]

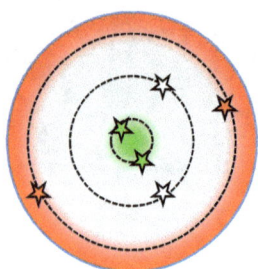

Figure 4

In this illustration the two feeling tokens represented by the green star symbols belong to the positive valence feeling type, the two feeling tokens represented by the red star symbols belong to the negative valence feeling type, and two feeling tokens represented by the white star symbols belong to the neutral valence feeling type.

Connecting the notion of valence to the previous discussion about global *qualia*, the outcome is that – if the hypothesis is correct – different feeling tokens (f_1, f_2 – say – the two green stars) supervening on different global neural state tokens (n_1, n_2) in response to different input token stimuli (tIS_1, tIS_2), despite having different *global* qualia could nonetheless share some phenomenal qualities. In other words, two objectively different scenarios (such as scenarios (8) and (9) in Table 11) could be related by virtue of a *partial phenomenal similarity* (e.g., a similarity in positive valence, indicated this as valence +) despite the

---

161 These illustrations are presented as pedagogic support and should be taken with a grain of salt. Undoubtedly it would be epistemologically problematic to actually map feeling tokens on the valence spectrum and in the global qualia space, but that is not the point. The only purpose of these illustrations is to help visualizing the idea that, despite their global phenomenal difference, different feeling tokens can share some (but not all) of their phenomenal qualities (i.e., in the specific case, valence).

fact that – on my assumption – there is no corresponding similarity to be found among the two scenarios in terms of global neural state tokens (n_1, n_2), input token stimuli (tIS_1, tIS_2), or even *global* qualia (f_1, f_2).

Table 11

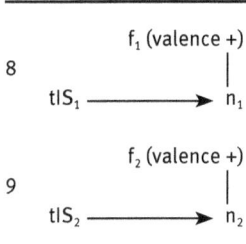

The bases for my hypothesis about the biological function of consciousness are now set. In the next section I explain in more detail how I think that the phenomenally perceived valence of any feeling token could explain how feeling could influence biological fitness.

7.3.5 The Reliability Of Valence Feelings

It seems sensible to claim that, in normal conditions, on average, feelings have a positive valence in situations that have a high objective probability of being fitness-enhancing, and – conversely – feelings have a negative valence in situations that have a high objective probability of being fitness-detrimental. By "situation" I mean the objective state of affairs to which a being belongs, not the input token stimulus perceived by the being. By saying that a situation has a "high objective probability" of being fitness-enhancing or fitness-detrimental I want to suggest that there is a fact of the matter as to which situational setups are potentially good or bad for any given being in a given context. To take an example, being engulfed in lava has a high objective probability of being fitness-detrimental for most (if not all) living beings. On the contrary, any hedonistic setup where appropriate food and mates surround a being and there is no potential danger is an input token stimulus that has a high objective probability of being fitness-enhancing for such a being. My claim is that, for example, feelings of satisfaction, joy, or excitement (i.e., with a positive valence) normally arise in situations that are objectively positive in terms of biological fitness (i.e., with a high biological efficacy) for the feeling-being in question, whereas feelings of pain, distress, or disgust (i.e., with a negative valence) normally arise in situations that are objectively negative in terms of biological fitness (i.e., with a low biological efficacy). Furthermore, on average, accentuated positive or negative valence of feeling tokens occur in borderline fitness situations, that is, situations in which one has a far above or below average fitness. For example, orgasms – which normally occur in particularly fitness enhancing situations – have a particularly positive valence, whereas pain due to severe wounds – which normally occurs in particularly fitness detrimental situations – has a particularly negative valence. This apparent correlation between valence feelings and objective biological fitness in most situations, if correct, may be interpreted as suggesting that, in normal conditions, on average, a feeling's valence reliably signals at the phenomenal level whether one is in a positive or negative situation with respect to biological fitness. My hypothesis about the biological function of feeling relies on assuming this to be the case.

My hypothesis continues by postulating a second caveat, namely that regardless of the fine-grained objective details of a given token input stimulus (i.e., regardless of the detailed information one has in any situation), feeling "roughly" good or bad in that situation *might be sufficient to tend to trigger selective behaviour* (e.g., appropriate reactions of approach or avoidance). More precisely, I suggest that feeling "roughly" good or bad in a situation might be sufficient to tend to trigger selective behaviour *even in the absence of explicit reasoning capacities of a fully-fledged mind*. This addendum is important because it suggests that valence feeling might be biologically useful even in the absence of higher-order cognitive capacities, and thus also that it might have been biologically useful before – phylogenetically speaking – the development of higher-order cognitive capacities. Indeed, it may be that the capacity of rational cognitive discrimination of concepts such as "good" and "bad" is possible in virtue of a more fundamental capacity for phenomenal discrimination between what *feels* good and what *feels* bad.[162]

Thus, to sum up so far, my hypothesis relies on the double assumption that (i) feeling beings *can* rely on a rough – but better than chance – pre-conceptual phenomenal individuation of situations that "feel good" or "feel bad", and that (ii) on average relying on valence feelings allows efficient behaviour in virtually any situation. In other words, the hypothesis relies first on the assumption that, on average, feeling good reliably and rightfully pre-conceptually informs the being that – say – the *status quo* ought to be preserved and that, conversely, on average feeling bad reliably and rightfully informs the system that – say – the

---

[162] This topic ought to be developed in depth, but it goes beyond the scope of the present work. This partially connects with what I introduced in section 4.4.

status quo ought to be altered. Secondly, the hypothesis relies on the assumption that valence feeling can causally contribute to guiding one's behaviour appropriately – i.e., trigger selective behaviour – in most situations. I back up these two assumptions in turn.

Let us assume, for the time being, the second assumption saying that valence feeling can cause selective behaviour (I come back to this in section 7.3.6). Biologically speaking, valence feelings are reliable guides to fitness-efficient behaviour only if they are appropriate – that is – if, on average, negative valence-feelings are especially salient in objectively fitness-detrimental (or potentially life-threatening) situations, whereas positive ones are especially salient when the current state of affairs is objectively fitness-beneficial. In other terms, valence feelings are reliable guides to fitness-efficient behaviour only if there is a better than chance probability that they correctly map the actual biological fitness of a being in a given situation. I suggested that this seems to be the case. But how could this be possible?

I think that a reasonable reply consists in saying that feeling valence in a way that encodes salient information about the world (e.g., positive valence in biologically beneficial circumstances, or negative valence in biologically detrimental circumstances) has been selected in the long run. On the contrary, feeling valence in a way that encodes inappropriate information triggering systematically counter-selective behaviour (e.g., positive valence in biologically detrimental circumstances, or negative valence in biologically beneficial circumstances) has been selected against in the long run. The reason for this is that, on average, beings that feel good in most circumstances in which their primary needs (e.g., alimentation, mates, shelter) are satisfied, and that feel bad in most circumstances in which such needs are not satisfied – where feeling like that reliably drives behaviour –, have higher chances of surviving until reproduction and fostering their genome than beings that feel good in most circumstances in which their primary needs are *not* satisfied and that feel bad in most circumstances in which such needs *are* satisfied. If this is correct, there would be a reasonable explanation for the apparent fact that, on average, negative valence-feelings are especially salient in fitness-detrimental (or potentially life-threatening) situations, whereas positive ones are especially salient when the current state of affairs is objectively fitness-beneficial[163].

One may still wonder how is it that feeling appropriately (i.e., in the "right" way) could be selected for, whereas feeling inappropriately (i.e., in the "wrong"

---

[163] It could be further claimed that feeling in a way that encodes neutral information or no information at all (i.e., neutral valence feeling) has been maintained because of economy.

way) could be selected against in the long run. An example should help in making this clear. Imagine a lion that is utterly sickened (i.e., has strongly negative valence feeling tokens) every time a gazelle (or any other possible lion-prey) is nearby. Let us suppose, in other words, that the lion feels revolted every time it sees, smells, tastes, or is around a gazelle. On the assumption that valence feelings can drive behaviour (or contribute to driving behaviour), it would be predictable that such a lion would avoid being around gazelles – never mind eating gazelles. This avoidance behaviour, clearly, is not fitness beneficial, since the lion needs to feed to survive. Let us further suppose that the lion also has negative valence feelings whenever lionesses are in heat, and that, because of this, the lion avoids fertile lionesses and does not reproduce. This behaviour, like the former, is not fitness beneficial. Finally, let us suppose that our poor lion friend has positive valence feelings whenever poachers with guns are around, and that, because of that, he is driven to get close to them (rather than avoiding them). Once again, the lion's behaviour is not fitness beneficial. Remember that biological efficacy consists in obtaining on average a positive outcome (a benefit) in the compromise between energy invested in performing a task and energy gained from it. The above-mentioned lion would not be able to react efficiently (quickly and well enough) to many possible configurations of token input stimuli, and therefore it would have a low biological efficacy. The morale of this example is that if valence feelings guide one's behaviour inappropriately (even if in less extreme cases than the one just mentioned) that could negatively impact one's biological fitness and in the long run lead to a negative selection process. Since it is by virtue of the subserving neurophysiological architecture that feelings have a specific global *qualia* (including valence), and since the subserving neurophysiological architectures that are selected in the long run are those of individuals that are biologically efficient, appropriate valence feeling is selected for in the long term, whereas inappropriate valence feeling is selected against. Indeed, if valence feelings guide one's behaviour appropriately (e.g., if the lion feels good when its preys and lionesses are around, and feels bad when poachers are around), on the assumption that feeling *can* positively impact behaviour and thus biological fitness, the subserving neurophysiological architecture on which appropriate valence feelings supervene is likely to be positively selected and – accordingly – the supervening appropriate valence feelings are likely to be preserved. This explanation does not come without problems, but I come back to that later.

It is important to highlight that evolutionary success is not measured on individual occurrences of a given behaviour, but rather on the general tendency to behave appropriately. Thus, it is fine to occasionally have inappropriate valence feelings as long as, on average, valence-based behaviour is successful. Because of this, the hypothesis that valence feelings are biologically efficient cannot be

state at t_2 might not be n_2, but rather a different token neural state n_3. That is, the additional valence-based information at t_1, by changing the token input stimulus at t_2, indirectly also contributes to determine which corresponding token global neural state is going to ensue at t_2. The causal chain should become apparent by confronting scenario 11 – in which valence enters the token global input stimulus – with scenario 12 – in which there is no valence feeling, indicated as $f_1 \varnothing$ (Table 13).

Table 13

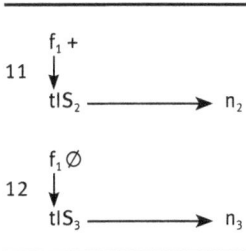

The difference in valence at t_1 between the two scenarios entails a difference in token global input stimulus at t_2 that in turn entails a difference in the token global neural states occurring at t_2 (i.e., n_2 and n_3). The fact of having a token global neural state – say n_2 – rather than another token global neural state – say n_3 – is what is directly causally relevant for behaviour. For example, we could imagine that n_2 causes an action a_2 whereas the different token global neural state n_3 causes a different action a_3 (see scenarios 12 and 13, Table 14).

Table 14

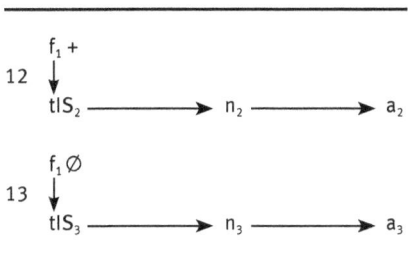

The upshot of the hypothesis is that even if (appropriate) valence feelings do not contribute in guiding behaviour by actively and directly *doing* something, they might nonetheless contribute in guiding behaviour passively and indirectly. More precisely, valence feelings might guide behaviour indirectly by making a difference in determining the token global input stimulus that is going to determine the ensuing token global neural state that is ultimately responsible for causing behaviour. In scenario 12 (Table 14), the positive valence of f_1+ indirectly contributes to the determination of the ensuing action a_2 by virtue of the fact that, counterfactually, if at time t_1 the global *qualia* had had a neutral valence ($f_1 \emptyset$) instead of the positive valence it has (f_1+), the action that would have ensued would have been a different one, namely a_3 instead of a_2. That is, the valence of the feeling at t_1 determines which scenario among 12 and 13 is actualised.

7.3.7 Learning to Behave Effectively

So far I have defended my hypothesis about the biological function of phenomenal consciousness by claiming that (i) on average, feeling beings *can* successfully rely on a rough – but better than chance – pre-conceptual phenomenal individuation of situations that "feel good" or "feel bad" (section 7.3.5), and by suggesting (ii) *how* valence feelings could impact behaviour (section 7.3.6). However, it might be contended that it is not clear how the *ontologically subjective* mode of existence of valence feelings could be important to prompt biologically efficient behaviour, and what is the biological advantage of being conscious *simpliciter*.

My answer to this question is best introduced by thinking about the ontogenetic development of a single being, and about how we learn from experience. A good starting point is Thorndike's law of effect, a basic principle of behavioural science positing experience and holding that the qualitative variable (i.e., satisfaction or discomfort) is crucial for the explanation of changes of behaviour (cf. Thorndike 1927)[164]. Here is Thorndike's original formulation:

> Of several responses made to the same situation, those which are accompanied or closely followed by satisfaction to the animal will, other things being equal, be more firmly connected with the situation, so that, when it recurs, they will be more likely to recur; those which are accompanied or closely followed by discomfort to the animal will, other things being equal, have their connections with that situation weakened, so that, when it recurs, they will be less likely to occur. The greater the satisfaction or discomfort, the greater the strengthening or weakening of the bond. (Thorndike 1911, p. 244)

---

[164] I am grateful to Markus Wild for addressing me toward this literature. See Wild (2008, pp. 37–38).

When Thorndike talks about different animal behavioural responses to the "same situation"[165], what he means is that a single being repeatedly facing the same fixed empirical setup paradigm will initially show various behavioural responses, some more biologically efficient than others. For example, a hungry cat repeatedly put in the same puzzle box will initially show various behavioural attempts to get to the food. Thorndike's law predicts that behavioural responses to that fixed empirical setup paradigm that in the past were "accompanied or closely followed by satisfaction" – that is, in our vocabulary, by a positive valence feeling – will tend to increase proportionally in time, whereas behavioural responses that in the past were "accompanied or closely followed by discomfort" – that is, in our vocabulary, by a negative valence feeling – will tend to decrease proportionally in time. For example, the law predicts that, the more the attempts, the more the cat's behavioural responses that in the past were accompanied by satisfaction will tend to increase proportionally. In other words, as the experiment proceeds, the responses by means of which the cat got to the food reward in the past will happen more frequently. Thorndike's law of effect suggests that there is a direct correlation between the qualitative variable of experience accompanying a given behavioural response (i.e., whether a being has a positive or negative valence feeling in a given situation), and the change in the frequency of such a behavioural response to the same situation over time.

A careful interpretation of Thorndike's law of effect has to highlight that even if the qualitative variable (i.e., satisfaction or discomfort) is crucial for the explanation of changes in the frequency of given behavioural responses, this does not necessarily entail that the quality of experience is *causally responsible* for the changes in the frequency of given behavioural responses. Nevertheless, the law of effect clearly supports the idea that valence feeling helps beings to progressively learn how to behave in a biologically efficient manner in a given situation by means of experience (i.e., by trial and error).

I want to suggest a strong variant of Thorndike's law of effect holding that in the long run, for any being, *by virtue* of valence feeling, beings tend to adopt biologically efficient behaviour in response to previously encountered situations by

---

[165] Thorndike's notion of "situation", intended as empirical setup paradigm, differs from my notion of "token input stimulus". The latter, remember, is defined as the *totality* of the information that a given beings accesses at a given moment – thus it is defined in relation to the being's informational affordances. Thorndike's notion of "situation", on the other hand, is not defined relative to the being, but rather in objective terms – as setup conditions for a given experiment. Even if the empirical setup paradigm remains unaltered across several experimental runs, the being's "token input stimuli" might differ from one run to the other because of other changes (e.g., changes in valence).

continuously reinforcing biologically efficient behavioural responses (i.e., those that felt good in the past), and weakening biologically inefficient behavioural responses (i.e., those that felt bad in the past). That is, I want to suggest that valence feelings are not just merely *correlated* with an increase in the frequency of given behavioural responses occurring in a given situation – as the law of effect suggests –, but rather, more strongly, I want to suggest that valence feeling could be the *cause* of the increase in the frequency of biologically efficient behaviour. Roughly, I want to suggest that what *explains* the increase in frequency of the cat's selective behaviour in a given recurring situation is the fact that the "right" sort of behavioural response feels more and more good *for* the cat as the experiment proceeds, whereas the "wrong" sorts of behavioral responses feel more and more bad *for* the cat. The way in which the cat feels, I claim, affects the way in which the cat behaves. Notice that in order to claim that the qualitative aspect of experience (i.e., valence feeling) is a potential *cause* of changes in behaviour, my variant of the law of effect holds that valence feeling *precedes* behaviour – rather than accompanying or following it as in Thorndike's original formulation. With this in mind, I formulate an *ad hoc* variant of Thorndike's law[166]:

> Of several behavioural responses occurring in response to an occurrence of a given situation, (i) the behavioural responses which are preceded by positive valence feelings will, other things being equal, be more firmly connected with the situation, so that, when an occurrence of that situations recurs, they will be more likely to recur, whereas (ii) the behavioural responses which are preceded by negative valence feelings will, other things being equal, have their connections with that situation weakened, so that, when an occurrence of that situation recurs, they will be less likely to occur.

This variant is stronger than Thorndike's original law because it does not only suggest that the quality of experience is closely connected to changes in the frequency of given behavioural responses; it suggests that the quality of previous experiences is *causally responsible* for the changes in the frequency of biologically efficient behavioural responses. If, as I suggested, positive valence tends to appropriately discriminate situations that are objectively fitness beneficial, and negative valence tends to appropriately discriminate situations that are objec-

---

[166] One difference between the original formulation and this variant is that the objective notion of "situation" is replaced by the being relative notion of "*type* input stimulus". I define the notion of type input stimulus as the set of all the token input stimuli that share a relevant amount of similarities. The subjects of Thorndike's experiments are confronted with invariant environmental setups, but their token input stimuli changes slightly every time because of previous experience (e.g., in addition to the present perception of the setup, there could a memory due to the past experiences).

tively fitness detrimental, then appropriate valence feeling tends in the long run to reinforce fitness-beneficial behaviours and weaken fitness detrimental behaviours, thereby permitting learning to behave ever more effectively by means of experience. My claim, in short, is that biologically efficient behaviour that follows fitting[167] valence feeling is more likely to recur in response to occurrences of the situation in the future.

But why is the *ontologically subjective* mode of existence of valence feelings important to prompt biologically efficient behaviour? I think that the ontologically subjective mode of existence of valence feeling is important because it allows beings to behave in a biologically efficient manner – i.e., better than chance – in situations that are objectively novel, but that feel subjectively like previously encountered situations. That is, valence feeling allows beings to behave in a biologically efficient manner in virtually any situation and in response to virtually any token input stimulus as long as there are subjectively salient similarities between that and a previously encountered situation.

Consider the example of a being S in a wildlife setting. S is in an objectively novel situation, i.e., a situation never encountered before. If the token input stimulus tIS_1 that S accesses at t_1 does not include any valence-based information – i.e., if S does not feel good or bad – the probability of S's behaviour being biologically effective will be – say – at chance level, somewhat like the cat at the beginning of the experiment. On the other hand, if the token input stimulus tIS_1 that S accesses at t_1 *does* include some valence-based information – i.e., if S *does* feel good or bad – the probability of S's behaviour being biologically effective will be above chance level. For example, even if the token input stimulus that the cat accesses is slightly objectively different at every trial, by virtue of the fact that, at the phenomenal level, objective differences are mitigated and objectively different situations can phenomenally appear as of the same type (see section 7.3.2), the cat can rely on a phenomenal reminiscence of subjectively alike situations to behave effectively. The interesting thing is that whether a token input stimulus tIS_1 feels good or not does not depend only on the present objective situation – it depends also on the being's previous experiences of that phenomenal type.

This goes beyond suggesting that qualitative variables (i.e., satisfaction or discomfort) are crucial for the explanation of behaviour, as Thorndike suggests. I hold that the behavioural "know-how" that is progressively acquired and refined through experience by responding more or less efficiently to given token input

---

[167] I use the term "fitting" both in the usual sense of "appropriate" and to remind the claim that valence feelings are "fitness beneficial" if positive in an objectively positive situation, or negative in an objectively negative situation.

stimuli can be applied to virtually any sort of token input stimulus that is relevantly similar – at the phenomenal level – to any of the token input stimuli already encountered. The token/type distinction of input stimuli is crucial in this respect. If a being is confronted with a novel *token* input stimulus that is also of an unknown phenomenal type (i.e., a type of feeling never encountered before), the probability of behaving efficiently will be – say – at chance level. On the other hand, if a being is confronted with a novel *token* input stimulus – as is very frequently the case –, but the latter is of a previously encountered feeling type, then the probability of behaving efficiently will be above chance level (depending on the case, even *significantly* above chance level). The more a being is experienced, the more the very fact of feeling good or bad in a (any) situation is likely to impact the token global input stimulus with the effect of increasing the probability of behaving in a biologically efficient manner (i.e., above chance level). If it is true that similarities in valence (i.e., feeling types) do not have a counterpart at the level of token global neural state similarities (i.e., global neural state types), and if it is true that valence is on average a reliable guide for behaviour, then it seems plausible that feeling might make a causal difference for behaviour and contribute to a being's biological fitness.

Summarizing, I began section 7.3 by saying that in the actual natural world, under the strong energy and time constraints imposed by life in natural settings, any living being has to be able to react efficiently (quickly and well enough) to as many possible token configurations of input stimuli as possible in order to have a high biological efficacy. I claimed that since beings constantly faced with complex and varying being-environment interactions face a nearly unlimited amount of possible token configurations of input stimuli they would benefit from a "rough" input recognition system (i.e., what I called a heuristic strategy) detecting patterns of similarities between input tokens and *quickly* producing *good enough* responses, thereby achieving a biologically satisfactory (albeit sub-optimal) result even in novel situations. Finally, I held that when heuristic neural information processing occurs in beings with complex integrated multi-modal sensory systems, the information about a given token state of affairs is automatically and involuntarily subjectively represented onto a weighted, unified "rough" feeling token. If – as I claim – some feeling tokens have a positive or negative valence, where, on average, valence reliably signals whether one is in a fitness beneficial or in a fitness detrimental situation, and if – as I also claim – valence feelings can indirectly prompt biologically efficient behaviour, then this might be a possible biological function of valence feelings.

One may contend that even if *valence* feelings appear to have a biological function, it does not follow that feelings *simpliciter* (i.e., global *qualia*) have a biological function. My reply is simply that valence does not occur independently of feelings since it is a qualitative property of some feeling tokens. Given that there would be no valence in the absence of feeling tokens, it is not strictly speaking the valence that has a biological function, but rather feelings. In other words, since it is only by virtue of feeling *simpliciter* that objectively different situations feel phenomenally alike and that it is possible to distinguish what feels good and what feels bad, the apparent biological function of valence is actually a biological function of feeling *simpliciter*.

My hypothesis about the biological function of valence feeling, as explained in section 7.3.4, does not originate within the standard emotion-centred valence literature. In fact, my hypothesis relies on the endorsement of a general valence theory whereby valence is a qualitative property of phenomenally conscious states of mind in general – rather than only a property of emotional states or emotion experiences[168]. I think that this holistic way of conceiving valence is promising and worth refining further as an alternative or complement to the more restrictive way in which valence is usually discussed in emotion literature.

7.3.8 Summary of the Hypothesis and Preliminary Conclusion

In a nutshell, my hypothesis is that the biological function of feeling consists in contributing to driving biologically efficient behaviour. For a being in an ever-changing natural setup the only way to maintain a high biological fitness is to distinguish in a better than chance way what is "good" from what is "bad" and act accordingly – that is, roughly, go for what is "good", and to avoid what is "bad". I suggest that feeling is an effective biological phenomenon that (i) in most situations reliably contributes to distinguish situations that are objectively "good" for the conscious being, from situations that are objectively "bad", even in the absence of conceptual reasoning[169], and that (ii) indirectly contributes to drive biologically efficient behaviour. If in a given situation you feel bad, even if

---

[168] On the biological function of the mind and emotions see Pinker (1997). On emotion and the function of consciousness see DeLancey (1996).

[169] Remember that throughout this chapter I have been considering valence as a first-order qualitative property. I do so because my goal is to answer the Natural Problem of consciousness, and doing so requires identifying a possible "primordial" biological function of phenomenal consciousness independent of reasoning and other high-order cognitive capacities. It would be very interesting to explore whether and how second-order desires, higher-order believes and the like might impact valence, and the extent in which valence is biologically functional for beings with a developed cognitive system, but this is not required to answer the Natural Problem.

you do not know *why* that is the case – indeed, even if you know nothing at all – you will naturally tend to escape that situation (or change the *status quo*).

According to my hypothesis it is a contingent fact that in the actual natural world "behaving and feeling" is more efficient than just "behaving without feeling" for living beings with a complex integrated multi-modal sensory system having to deal with ever changing complex environments. On these grounds, it is reasonable to suppose that those species that are faced with complex and varying being-environment interactions (like humans) feel[170]. If it is correct to affirm that feeling is a natural phenomenon that can contribute to yield biologically efficient behaviour (i.e., behaviour increasing the biological fitness of beings living in complex environments), then this hypothesis is a viable candidate to explain why some beings presently feel, and – as a consequence – also a viable explanatory answer to the Natural Problem of Consciousness.

Either something along these lines is the case – i.e., either feeling can *somehow* influence biological fitness –, or feeling is epiphenomenal. My top-down hypothesis on the possible biological function of feeling is far from being uncontroversial and I do not claim that it is conclusive. However, as I have argued in section 6.4.2, an attempt to solve the Natural Problem should strive to offer an *explanatory satisfactory* theory of why feeling presently exists. Regardless of whether or not my hypothesis is ultimately correct, I think that the argumentative strategy I have advocated has at least the merit of offering a philosophically engaged and explanatorily rich alternative to epiphenomenalism as a viable solution to the Natural Problem.

The main goal of the present work has been achieved. I dedicate chapter 8 to some final considerations on the accounts of causation that might be compatible with my hypothesis, a note on how to support my hypothesis empirically, and the conclusion.

---

[170] According to my hypothesis it is less clear whether species faced with relatively simple and stable being-environment interactions also feel, since – arguably – in those conditions a set of rigid input-output instructions might be sufficient to foster selective behaviour.

8 Causation and the Conscious Mind

I have claimed that an explanatory answer to the Natural Problem of Consciousness has to concede that consciousness has a biological function, i.e., that it can play a causal role that can affect biological fitness. The hypothesis about valence that I advanced in chapter 7 is developed precisely on this basic assumption. I suggested how feeling could prompt fitness beneficial behaviour by playing a causal role that nothing else could play in a similarly efficient manner. So far, however, I have been relying on an unspecified notion of causation. In order to clarify the hypothesis, in section 8.1 I briefly consider which account of causation might be compatible with my hypothesis in claiming that feeling could play a causal role in influencing biological fitness. On this basis, I question whether we could empirically test whether valence feeling and biologically efficient behaviour are causally related (section 8.2). After these last remarks, I present my conclusion (section 8.3).

8.1 Theories of Causation

Causation is a complex self-standing object of philosophical enquiry, and this is not the place to explore what causation in general consists in, never mind taking a stance on what kind of causation is required to make sense of causal connections in the actual natural world. My goal, thus, is limited to raising a few specific remarks aimed at suggesting which conception of causation might be compatible with my hypothesis about the biological function of feelings. I introduce and briefly discuss three accounts of causation: causation as constant conjunction (section 8.1.1), causation as counterfactual dependence (section 8.1.2), and causation as transfer of physical quantities (section 8.1.3)[171]. I then summarize the findings, emphasizing that my hypothesis about the biological function of valence feelings does not require a commitment to more than a counterfactual theory of causation (section 8.1.4).

---

[171] This section follows the setup of the introduction to causation by Mumford and Anjum (2013). For more on causation see Beebee, Hitchcock and Menzies (2009), and Sosa and Tooley (1993). On contemporary Humeanism see Beebee (2006); Loewer (2007a); Cohen and Callender (2009). I am grateful to Michael Esfeld for comments and suggestions for improvement on an earlier draft of this chapter.

8.1.1 Causation as Constant Conjunction

In presenting different accounts of causation, a good place to start from is the regularity account of causation offered by the empiricist David Hume. Hume, who was interested in explaining what we think of as a cause, remarked that we can observe a series of events following one another, but we never observe the causal connections themselves (1739, Book I, Part III, Section VI; see also Hume 1748). What we do when we talk about causes, or about something causing something else, is infer the presence of a causal connection from an observed regular succession of events. This Humean observation does not necessarily entail that causation does not exist, but it highlights that our commonsensical idea of causation (i.e., causation as commonly understood by philosophers) cannot directly be explained by the manifest existence of something in the world[172]. In Hume's view, the reason why we think that one kind of event causes another is that we observe *regularities*. For example, the event C (cause) "a billiard ball hitting a stationary ball" is always followed by the event E (effect) "the second ball starting to move". According to Hume the observation of such regularities between a cause and an effect – also known as "constant conjunctions" – explains the fact that we come to think that, for example, the first ball hitting the second *caused* the second ball to move, even though – as a matter of fact – we only observed a sequence of separate events C and E.

> All events seem entirely loose and separate. One event follows another; but we never can observe any tie between them. They seem conjoined, but never connected. (Hume 1748, section VII, part 2).

In sum, Hume and later-day Humeans such as David Lewis suggest that we assume the existence of causation on the basis of the observation of regular patterns of conjunction. Interestingly, even if we might think that there is a "real" cause responsible for the regularity of the conjunctions, for Humeans no such thing is required: any apparent "real cause" could ultimately be explained and resolved in term of further regularities. Hume grants that it is natural to infer from previously observed regularities (E following C) given expectations about what will happen in future unobserved cases of – say – balls hitting other balls. However, he claims that such inductive inferences – no matter how popular and successful – are not rationally justified.

Hume, however, grants that the observation of a constant conjunction of two events C and E is not alone sufficient to claim that there is a causal connection

---

[172] See also Russell (1913).

between them, nor is it sufficient to explain the notion of cause. What is needed in addition to that, according to him, is a commitment to temporal priority – i.e., the idea that causes must be temporally antecedent to their effects –, and spatial contiguity – i.e., the idea that causes must be spatially adjacent to the effect. Temporal priority allows the explanation for asymmetry and directionality of causation, highlighting which event is the cause and which is the effect, by suggesting that the constant conjunction of an event C with an event E (in *that* sequence) excludes the constant conjunction of event E with event C (in *that* sequence) – something that could not be excluded by constant conjunction only. For example, hitting a piano key causes a sound by virtue of the fact that the constant conjunction of these two events always comes in this specific order: first the key hitting (the cause), then the sound (the effect), and not *vice versa*. If it were the case that sometimes the sound precedes the key hitting, then the constant conjunction would not indicate a causal connection between the two events. In addition to this, Hume suggests that spatial contiguity is also part of our notion of cause, since causal processes do not occur over a distance, even though there might be causal chains of contiguous events creating mediated indirect effects at a distance[173]. For example, hitting the piano key causes the sound only if the piano key is spatially contiguous – or adequately connected by a spatially contiguous mechanism – to the strings whose vibration produces the sound. If the sound comes from another piano than mine – one that is not spatially contiguous to the keyboard of my piano –, it is not *my* key hitting that is causally responsible for that sound, but rather – probably – a different key hitting, one that is spatially contiguous to that piano. According to this view, thus, causation requires constant conjunction, temporal priority, and spatial contiguity between two events.

Hume's proposal is appealing, but there is a tension between the requirements of temporal priority and spatial contiguity. Take the example of causation in the case of billiards. No causation occurs until the first ball (the cause) is spatially contiguous to the second (the effect), since – according to spatial contiguity – it is only when the two balls touch that the first ball can cause the second ball to move. Moreover, as soon as the two balls are spatially apart, the causation process is over. Given that the spatial contiguity that Hume takes to be required for causation lasts for a distinct moment in time (or time period), the cause does not however, strictly speaking, occur prior to the effect, but is rather simultaneous with the effect. That is, the causing event has to be somehow temporally

---

[173] This claim might be challenged on the basis of quantum entanglement, which suggests the possibility of action over a distance and, possibly, of causation in the absence of spatial contiguity. Hume, of course, could not have thought of this.

superimposed with the caused event. This seems to challenge the temporal priority claim that causes can be distinguished from effects because they always precede them. One way out of this impasse would be to accept the Kantian idea that simultaneous causation is possible (even if not all causal processes need be simultaneous), but then we might have to accept, pace Hume, that cause and effect might be indistinguishable.

An even more pressing problem for the Humean regularity view consists, however, in distinguishing accidental regularities – coincidences – from genuine lawlike causal regularities. In fact, Hume would have to grant that *any* contingent constant conjunction between any two events C and E is a case where C and E are causally connected. This is a problem because we can easily find examples in which two events are regularly conjoined (maybe just a few times), where the requirements of temporal priority and spatial contiguity are met, but where this happens only *accidentally*. The Humean has to acknowledge that these accidental regularities are indistinguishable from regularities that – according to common sense – are due to supposedly genuine lawlike causal conjunctions. For example, imagine it happened twice in the past that you dropped your phone on the ground just after a leaf fell on your shoulder. It is tempting to say that the leaf falling on your shoulder is not the genuine cause of you dropping the phone, but rather just a coincidence. After all, it is doubtful that every time a leaf falls on your shoulder in the future, you will drop your phone. According to the Humean, however, the only thing we know for sure in this case is that in the past there has been a constant conjunction between these two events (so far, every time a leaf fell on your shoulder you dropped the phone!). On the basis of this observation alone, until proven wrong, there is no reason to suppose that the leaf falling on your shoulder is not causally connected with the dropping of the phone. In sum, since we cannot observe causation directly and since we can only base our causal judgements on past conjunctions of events, there is no way to rationally distinguish supposedly genuine lawlike causal connections from mere accidental ones.

Besides regularity, temporal priority, and spatial contiguity another important criterion that is discussed in theories of causation is the *necessity* of causes, the idea that something is a cause if it necessitates or strictly compels a specific effect. We can find this idea already in Baruch Spinoza, saying, "from a given definite cause an effect necessarily follows" (1677, I, axiom III). This is a "realist" view of causation because it takes causes to be a real thing that is sufficient to genuinely produce or bring about their effects, rather than just being contingently regularly connected to effects as suggested by Hume. The necessity condition has the great advantage of avoiding the Humean problem of distinguishing accidental contingent relations between events (i.e., coincident regularities) from genuinely causal relations between events. In fact, for someone who is a necessitarian realist regarding causation, something is an effect only if it occurs *because of* a necessary cause and not – as Hume says – by being contingently regularly conjunct with it in an appropriate manner. According to the necessitarian view, for example, the leaf falling on your shoulder could be excluded as a cause of your dropping the phone because, regardless of the contingent accidental correlations of the past, it is not *necessary* for the correlation to hold (i.e., it is possible for a leaf to fall on your shoulder without you dropping the phone). If causation is real, as the anti-Humean claims, this would also explain why we observe regularities in nature, the apparent contingency of regularities being a symptom of real lawlike necessary causal relations.

Not all realist (non-Humean) accounts of causation rely on the strong requirement of metaphysical necessity. Indeed, there are several theories lying between the extremes of Humean regularities and metaphysical necessity. Anscombe (1971), just to cite an interesting example, suggested for example an indeterministic notion of causal production that preserves the idea that causes genuinely produce specific effects, but that does not rely on the necessity requirement. Anscombe based her claim on the physical notion of indeterministic causation suggesting that there are objective probabilities (that is, probabilities that are not entirely random), for something to happen[174]. For example, a given radioactive particle might have a high propensity (i.e., a natural inclination or tendency) to decay within a certain time, and there might be an objective probability for that to happen. This means that out of many such particles most of them will decay within that time-lapse, but there is no necessity as to which individual particle will decay and which will not. Anscombe's idea is that even though the effect might not be strictly necessitated by a cause, the cause is responsible for the fact that the effect has a high propensity to occur. Following this view, a cause does not necessarily ensure an effect, but it raises the objective probability of the effect occurring, and in this sense it still genuinely causes the effect.

Determining whether or not real causation is required to make sense of causal connections in the actual world is an interesting problem in the metaphysics of causation depending on several important issues, such as whether or not the Humean can distinguish between accidental and lawlike regularities, and on the nature of the *relata* of causal relations. However, given that this exceeds by far the goals of my dissertation, I remain neutral as to what is the best way to make sense of causal connections in the actual world. Nevertheless, both the Humean and the realist accounts of causation sketched above rely upon the general idea

---

[174] Anscombe's view no longer represents the state of the art. For a wider picture of causation and probability see Mellor (1971) and Suárez (2011).

of causation as constant conjunction. Are these accounts *prima facie* compatible with my hypothesis about the biological function of valence feeling?

If causation is better defined as mere constant conjunction between events, then valence feelings could count as the cause of biologically efficient behaviours if they are *contingently regularly conjoined* with them in the appropriate way. I have claimed that, in normal circumstances, valence feelings occur temporally prior to given biologically efficient behaviours, so this requirement is met. However, according to my hypothesis, valence feeling contributes in determining the global token input stimulus, but it does not guarantee that this always leads to biologically efficient behaviour (i.e., it is not the case that every time a being has a valence feeling it behaves in a biologically efficient manner), thus it is arguable that valence feeling is not *regularly* conjoined with biologically efficient behaviour. Furthermore, if the "appropriate" regular conjunction also requires *spatial* contiguity between valence feelings – which have an ontologically subjective mode of existence – and biologically efficient behaviour, this would pose a problem. In fact, it is not clear how a valence feeling – that exists in a non-spatial phenomenal dimension – could be spatially contiguous with any sort of behaviour. All things considered, thus, it seems that my hypothesis about the biological function of valence feeling is incompatible with a Humean account of causation as mere regular conjunction between events.

If causation is better defined as "real", on the other hand, my hypothesis should be read as claiming that valence feelings are not just contingently regularly conjoined with given biologically efficient behaviours – as the Humean version would have it –, but that they would be the "real" *cause* of given biologically efficient behaviours, in the sense that the existence of valence feelings is what genuinely produces or brings about given biologically efficient behaviours, or – at least – an increase in the objective possibility of a given biologically efficient behaviour to occur. *Prima facie*, my hypothesis about the biological function of valence feeling seems to be compatible with this sort of view, since what I claim is that valence feeling can influence a being's subsequent global token input stimulus, and thereby – indirectly – contribute in determining which token global neural state and ensuing behaviour is going to ensue.

8.1.2 Causation as Counterfactual Dependence

A refined regularity theory of causation addressing the problem of distinguishing accidental from genuine lawlike regularities is the one relying on the notion of counterfactual dependence between events, namely the idea that an event C is the cause of an event E if the absence of C would have entailed the absence of

E (i.e., the occurrence of E counterfactually depends on the occurrence of C)[175]. For example, we might describe the causal relation between the hitting of the piano key and the sound by saying that hitting the key is a difference-maker in the absence of which the sound would not have occurred. Similarly, the match striking can be described as the cause of the match being ignited by virtue of the fact that the match striking is the difference-maker in the absence of which the match would not have been ignited. More in general, a counterfactual dependence test of causation consists in questioning what would have happened if, other things being equal, – contrary to the facts – a supposedly causal event had not occurred. For example, imagine a law trial in which it has to be established whether Wilhelm Tell shooting his crossbow is the cause of the death of the bailiff Gessler. A counterfactual dependence theorist would say that Tell's action is the cause of Gessler's death if in the same situation – conditionally – had not Tell shot his crossbow, other things being equal, Gessler would not have died. If, for example, it could be proven that despite Tell not shooting his crossbow, other things being equal, Gessler would have died anyway – say, because of a heart attack –, then Tell's action could be discarded as cause of Gessler's death. The counterfactual dependence theory of causation relies on the Humean idea that causation is a contingent relation between distinct events, but differs from the simple regularity view (section 8.1.1) by casting the relation in terms of counterfactual dependence of events rather than in terms of mere constant conjunction between events.

A realist about causation would contest the counterfactual dependence claim according to which causation is counterfactual dependence between events, or – in other words – that counterfactual dependence is that in virtue of which events are causally related. The realist would say that this account gets the order of things wrong: counterfactual dependence of events is just a symptom, and it is explained by the fact that those events are *really* causally related. For example, the counterfactual claim that if Tell had not shot his crossbow Gessler would not have died is valid precisely because Tell's shooting was the real cause of Gessler's death in the actual world.

The counterfactual dependence thesis could be attacked by presenting cases of causation without counterfactual dependence or, vice versa, counterfactual dependence without causation. A case of causation without counterfactual dependence could be a case in which an effect is overdetermined by two causes.

---

[175] Most contemporary Humeans defend a counterfactual dependence theory rather than the simple regularity theory (but see Baumgartner 2008 for a reassessment of simple regularity theories). For more on counterfactual dependence see Collins et al. (2004).

For example, supposedly, Gessler's death could be caused simultaneously (overdetermined) by both Tell's shooting and a heart attack. In this case, since Tell not shooting would still have been counterfactually followed by Gessler's death (because of the heart attack), and the absence of the heart attack would still have been followed counterfactually by Gessler's death (because of Tell's arrow), neither Tell's shooting nor the heart attack, taken independently, count as causes of Gessler's death, according to the counterfactual dependence thesis. Another worry comes from the opposite case of counterfactual dependence without causation: the case of *sine qua non* necessary conditions. Gessler's death counterfactually depends on many things, including Wilhelm Tell's parents having intercourse and procreating Tell. In fact, if the latter event had not happened, Tell would never have been in a position to kill Gessler. Even though such an event is a necessary condition without which Tell would not have killed Gessler, that is not what we are looking for when asking what caused Gessler's death. Or, at least, there is a sense in which Tell's shooting is more crucial to Gessler's death than Wilhelm's parents having intercourse. There are possible replies to these worries, but – leaving them aside – this suggests that also counterfactual dependence between two events may be questioned as a satisfactory account of causation.

As I said before, I remain neutral as to what is the best way to make sense of causal connections in the actual world. However, I am interested in knowing whether my hypothesis saying that valence feeling is causally related with biologically efficient behaviour is compatible with a counterfactual dependence theory of causation. According to this theory of causation, valence feeling would be causally related with biologically efficient behaviour if biologically efficient behaviour counterfactually depends on valence feelings – i.e., if, other things being equal, the absence of valence feeling would have entailed a difference in biologically efficient behaviour. This seems to be compatible with my hypothesis, since all I need to claim is that valence feeling makes an indirect difference in determining the behaviour that ensues.

8.1.3 Causation as Transfer of Physical Quantities

Besides considering causation as constant conjunction between events (section 8.1.1) or as counterfactual dependence between events (section 8.1.2), one could think of causation as the transfer of conserved physical quantities such as energy, momentum, charge, and so on. According to transfer theories of causation (see Dowe 2000; Kistler 2006), causal processes consist in the transfer of physical quantities from causes – that lose a given physical quantity – to effects – that acquire a given physical quantity –. This sort of theory, besides not being independently unproblematic, is clearly not compatible with my hypothesis about how valence feelings causes biologically efficient behaviour. The reason for this is quite obvious: since I take valence feelings to be an ontologically subjective property of experience, it would be puzzling to claim that a valence feeling causes selective behaviours by transferring a given conserved physical quantity (an ontologically objective property) to something else. Endorsing a transfer theory of causation would lead us directly to the classical problem of mental causation intended as the problem of explaining how the mental could causally interact with the physical given the causal closure of the physical world. In the absence of strong independent reasons to assume this theory of causation over the alternative ones, I thus leave it without further ado.

8.1.4 Viable Theories of Causation

The outcome of this brief preliminary survey of some theories of causation is quite reassuring. Claiming that valence feeling causes selective behaviour by transferring a conserved physical quantity (section 8.1.3) would surely be controversial, but there are two alternative theories of causation – a form of realism about causation (section 8.1.1) and causation as counterfactual dependence (section 8.1.2) – that appear to offer viable solutions. In particular, my hypothesis could hold either if (i) valence feeling has a biological function by virtue of being "really" – directly or indirectly – causally responsible for the production of biologically efficient behaviour, or if (ii) in the absence of global qualia of a give type (i.e., valence feelings), other things being equal, there would not be similarly biologically efficient behaviour. Both these theories of causation are in principle compatible with my hypothesis that valence feelings can cause biologically efficient behaviour. Moreover, my hypothesis about the biological function of valence feelings does not *require* a commitment *to more* than a counterfactual theory of causation (such as, for example, a commitment to necessary causal connections).

8.2 Testing the Consciousness-Behaviour Causal Relation

In order to support the top-down hypothesis that valence feeling is somehow *causally related* to biologically efficient behaviour, it would be good to have some sort of empirical proof of the fact that, in the actual natural world, the absence of valence feeling, other things being equal, entails a difference in the biological efficacy of behaviour. We cannot apply the contrastive method adopted by bottom-up approaches to identify functional differences of particular transitively conscious mental states (see sections 1.3, 7.1, 7.2) because valence feeling is an intransitive state of mind. However, we could try to test whether there is at least a counterfactual dependence causal relation between valence feeling and biological effective behaviour by adopting a standard medical procedure – a randomized controlled trial (henceforth RCT).

The RCT consists in observing two groups of people formed at random from a large sample of test subjects[176]. In standard RCT, one group (the "treatment" group) is given – say – a trial drug that needs to be tested to see what its effects are, whereas the other group (the "placebo" group) is given a placebo. If the initial sample is large and the subdivision in groups is randomized, then the two groups will be so similar as to be almost alike in all factors that could interfere with the result of an experiment. In some sense, we could consider these two groups of subjects as exemplifying twice the same group of people (i.e., as if there was a single group of people and its counterpart in another possible world – but not requiring the metaphysical postulation of other substantive worlds as in Lewis' modal realism; see Lewis 1986). None of the test subjects know whether they are receiving the actual drug or just a placebo. If there is an observed statistically significant difference between the results of the two groups after the trial, given that the only objective difference between two groups consists in the taking of the drug or the placebo, it is rationally justified to conclude that the result suggests that there is a statistically better recovery rate in one or the other scenario. If, for example, in a given experiment the treatment group shows a statistically better recovery rate from a given disease than the placebo group, then we have an empirically based justification to claim that taking such a drug ensures a statistically better recovery rate than not taking the drug. We can claim this because the placebo group exemplifies and proves the counterfactual claim that, had the subjects of the treatment group not taken the drug, they would have had a poorer recovery rate. Notice that all you can observe is that there is a "statistically better" recovery rate if you take the drug. Claiming this is more conservative than claiming that the recovery rate is better *because* of the drug, or – the other way around – that the drug is the "real" cause of the better recovery rate. Since

---

[176] The non-randomized version of this method has its roots in John Stuart Mill's method of difference (1843, vol.1): "If an instance in which the phenomenon under investigation occurs, and an instance in which it does not occur, have every circumstance in common save one, that one occurring only in the former; the circumstance in which alone the two instances differ, is the effect, or the cause, or an indispensable part of the cause, of the phenomenon". The idea of adding randomization to the basic experimental design in order to remove selective or biasing influences is due to the statistician Ronald Fisher (1935).

it is possible to explain all the facts observed by means of RCT without having to assume a realist picture of causation, and since a realist picture of causation gives a causal interpretation of the data that might or might not be true, it is fair to interpret RCT as only suggesting a counterfactual dependence between events.

This counterfactual dependence test of causation is taken seriously in science, but can it be adopted as a test to verify whether valence feeling (or any kind of ontologically subjective phenomenon, for that matter) is causally counterfactually related with biologically efficient behaviour? We would need two similar groups of subjects, one of which has valence feelings – call it "normal group" – and one of which – other things being equal – does not have valence feelings – "abnormal group" –. If in a trial we observed a statistically significant difference in the biological efficacy of behaviour between the two groups – i.e., if the behaviour of normal group subjects is – say – statistically significantly more objectively biologically efficient than the behaviour of the abnormal group, then we would have empirical support for the claim that valence feeling ensures a statistically more biologically efficient behaviour than non-valence feeling. We could claim this because the abnormal group would exemplify and prove the counterfactual claim that, had the subjects of the normal group not had valence-feelings, they would have had less efficient biological behaviour. Ideally, this seems to be a good test. In practice, however, there would be several problems with the implementation and interpretation of a RCT for valence feeling.

A minor technical/methodological problem is that the subjects of the RCT for valence feeling would not all remain oblivious as to the group to which they belong until the end of the trial (compare with the normal case in which you do not know if you took the real drug or just a placebo). In fact, because of the very nature of the trial, the presence of valence feelings will be phenomenologically manifest to the subjects. The subjects of the normal group might know that they are not in the abnormal group because they still have valence feelings, and it is impossible to "hide" them (we might wonder whether the opposite is the case for the abnormal group). Knowledge of being in a specific group might – at least in principle – affect the behaviour of the subject and, by extension, compromise the results of the RCT.

A second and more important technical/methodological problem is the strict implementation of the *ceteris paribus* clause. Given my naturalist assumptions, even if it were possible to have "normal" subjects that have valence feelings, I do not think it would be nomologically possible to have "abnormal" subjects that do not have valence feelings *all other things being equal*. In fact, according to my hypothesis, because of supervenience and mind-body dependence, in the actual natural world in order to suspend valence feeling it would be necessary to alter the token global neural state on which they supervene. That is, since there can be no difference in valence feeling without a difference in their subserving base, it would be nomologically impossible to have subjects lacking valence feeling, but having the same token global neural states as the "normal" valence feeling subjects. If we ignore this fact, and pursue the counterfactual dependence test despite the double difference (of valence feeling *and* subserving base) between the two groups of subjects, this would lead to the troublesome problem of how to *interpret* the RCT results in the case of valence feeling.

In the case of the RCT drug test, the conclusive observation is straightforward: if there is a statistically significant difference between the two groups, then we can claim that – say – taking the drug ensures a statistically better recovery rate than not taking the drug. In the study of valence feeling, however, if there is a statistically significant difference between the two groups, then we should claim that – say – the statistically better selective behaviour counterfactually depends on the *conjunction* of a given physical substrate and the valence feeling supervening on it (a case of causal overdetermination). That is, even if we could empirically prove by means of an RCT that there is a counterfactual dependence between valence feeling and – say – increased biologically efficient behaviour, on the assumption that valence feeling contingently always co-occurs with and supervenes on a given token global neural state, we could not conclude *only* that what ensures a statistically better biologically efficient behaviour is the having of valence feelings. We would have to conclude that there is a counterfactual dependence between, on the one hand, valence feeling *and* its neural subserving basis, and, on the other hand, increased biologically efficient behaviour.

If this is right, an RCT for valence feeling would lead to a counterfactual dependence claim where distinguishing valence feeling and its subserving base as "apparent" and "real" cause of the biologically efficient behaviour is not possible, but neither is it required. Interestingly, in fact, for someone with no realist concerns about causation there would be no *a priori* reasons to prefer one counterfactual dependence claim to the other (e.g., to say that the counterfactual dependence between subserving basis and biologically efficient behaviour is more basic or fundamental than counterfactual dependence between valence feeling and biologically efficient behaviour, or vice versa). An RCT for valence feeling could only show that there is a counterfactual dependence between valence feeling *and* their subserving basis, and biologically efficient behaviour.

The realist about causation might contend that even if biologically efficient behaviour were found to be counterfactually dependent on the presence of both valence feeling *and* its subserving basis, the statistically better biologically efficient behaviour (supposedly) observed in the "normal" group by means of the RCT would only be *causally explained* by *one* of them, namely by the only "real" cause of behaviour. The reason for this claim would be the principle of absence of regular overdetermination saying that a physical effect (e.g., a behaviour) cannot have both a complete and sufficient physical cause (say, a given neural state) *and* a complete and sufficient mental cause that are equally regularly causally relevant for the effect in question (see section 2.1.5). Within a non-reductive physicalist framework such as the one I have conditionally endorsed, on the assumption of Kim's causal exclusion argument, this would come down to claim that the "real" cause of biologically efficient behaviour is the physical one (i.e., the neural subserving basis), and that valence feeling would therefore be excluded from the "real" causal story[177]. Notice, however, that the RCT alone does not support this realist conclusion. That is, the RCT alone does not prove that the statistically better biologically efficient behaviour observed in the "normal group" (i.e., the feeling group) over the "abnormal group" (i.e., the non-feeling group) is due, caused, or explained solely by a difference in neural substrate.

Leaving aside the metaphysical problem of mental causation, which is not of our concern here, the upshot of this is that the widespread and empirically acceptable RCT is not suited to test the counterfactual dependence hypothesis that in the actual natural world the absence of valence feeling, *all other things being equal*, entails a statistically significant difference in the biological efficacy of behaviour. An RCT could only support the explanatorily weaker counterfactual dependence claim that a difference in valence feeling *and* their subserving basis, other things being equal, entails a statistically significant difference in the biological efficacy of behaviour. This of course would not exclude that my hypothesis that valence feeling can cause biologically efficient behaviour *could* be correct, but neither would it straightforwardly support my hypothesis.

The problem of empirically assessing whether phenomenal consciousness has a function and what that function is, is not a problem only for my top-down hypothesis about intransitive valence feeling, but – as we have seen extensively in section 7.2 – also for bottom-up approaches to transitive conscious mental states. Thus, regardless of whether we are interested in state consciousness or creature consciousness, and regardless of the notion of causation we endorse, arguably the most pressing problem in the scientific study of consciousness remains that of developing a reliable methodology allowing us to gather satisfactory epistem-

---

[177] An interesting line of response to Kim's causal exclusion problem consists in suggesting that overdetermination is a problem only for "realist" accounts of causation, whereas it is harmless for "non-productive" accounts of causation such as counterfactual dependence (see Loewer 2001, 2002, 2007b, and Russo 2013). See Kim (2005, 2007) for an opposite view, holding that a counterfactual dependence theory of mental causation is insufficient to explain mental efficacy.

ically objective data that could unquestionably indicate the *exclusive* function of phenomenal consciousness.

8.3 A Conservative Conclusion to the Natural Problem

The main goal of the present dissertation was to introduce and unpack the Natural Problem of Consciousness. Why are there presently conscious beings at all? How can we explain the fact that, for example, humans and other animals have turned out to be conscious, rather than evolving as physical systems with no conscious mental life? I have suggested that two lines of argument can be defended. On the one hand, it is possible to defend epiphenomenalism and claim that feeling does not matter causally, in which case its present existence is best explained by the fact that feeling is linked to whatever does the causal job of producing selective behaviour in such a close manner that getting rid of feeling would require getting rid of its subserving base, which would be biologically inconvenient and detrimental. On the other hand, it is possible to claim that feeling matters causally and that its present existence is best explained precisely by the fact that feeling is biologically efficient. The latter line of argument is more ambitious, given that it has to solve several problems along the way, but – if accurate – it has the great advantage of providing a more explanatory satisfactory reply to the Natural Problem. In the concluding part of the dissertation I advanced and suggested one way to defend the latter line of argument. I did so by developing a top-down hypothesis suggesting how valence feeling could contribute to biologically efficient behaviour.

I do not claim that my hypothesis is a conclusive answer to the Natural Problem. Moreover, certainly, my hypothesis is not an answer to the Hard Problem of consciousness, which is a different problem that – as I have said explicitly – I have not tried to answer. Nevertheless, I believe that this research work has produced several philosophically valuable results along the way. First, it has displayed and outlined the Natural Problem, contrasting it and connecting it with other important problems of consciousness. Second, it has displayed the lines of argument one might want to pursue to answer this problem. Third, it has raised and mapped several issues connected with the definition, conception, and study of consciousness, producing numerous distinctions and clarifications. Finally, – hopefully – this work has suggested numerous lines of analysis and further research, contributing to the literature in particular by raising awareness as to the interest in (i) taking seriously the diachronic questions about consciousness, (ii) thinking about consciousness as an ontologically subjective biological phenomenon, and (iii) considering consciousness intransitively, as a property of some global states of mind of some beings. All in all, even if my hypothesis about the biological function of valence feeling is not to go undisputed, this work has framed the problem, and this is a positive philosophical result.

Appendix: Objections and Replies

In this appendix I consider a selection of questions and concerns that have emerged in various forms and in different contexts, in particular during my doctoral thesis defense and public presentations of my views on the Natural Problem. The first part of the appendix deals with methodological issues regarding the organization, scope and achievement of the work (1–4). The second part of the appendix deals with issues connected with the nature of creature consciousness and global qualia (5–7). The third part of the appendix deals with metaphysical issues and the problem of mental causation (8–12).

I am very grateful to all those who raised these issues – and particularly to Prof. Markus Wild and Prof. Michael Esfeld – for leading me to reflect further on these matters and allowing me to better clarify my motives and position.

[1] The presentation of the problem of accounting for consciousness, considering that this is a well-known problem in philosophy, is excessively long.

The problem of accounting for consciousness lies at the center of a vast and highly disputed debate in Philosophy and in the science of consciousness. It is true that the problem of accounting for consciousness is well known, but not all the aspects of this problem are equally well known by philosophers and scientists working on consciousness. It is fundamental, for the sake of clarity, to set aside important problems that are only marginal to the argumentative structure of the present work in order to highlight the originality of the contribution, whilst also situating the latter within the wider context of the interdisciplinary debate on consciousness. In this sense, although I agree that the presentation of the problem is quite long, I think this is justified by the need to frame the Natural Problem within a consistent body of interdisciplinary work and thereby set solid foundations for its analysis.

[2] Even though it is clearly stated that the main hypothesis developed in response to the Natural Problem is far from being uncontroversial, there is no in-depth discussion of possible objections to it. This could help in providing a clearer contour to the hypothesis.

This criticism is fair, in the sense that even though I do raise and briefly consider some directly pertinent objections – that is, objections that, by assuming what I assume, could be raised to question my hypothesis – it would have been indeed profitable to consider and discuss more in depth possible objections to my hypothesis in order to clarify it further. Nevertheless there are reasons why I have not done so.

The first reason for this is that the main goal of the dissertation is to introduce and unpack the Natural Problem, and not to develop a full-fledged answer to the Natural Problem. This explains both the long introduction and the treatment of my hypothesis as theoretical hypothesis. Even though chapter 7 presents the main hypothesis, the philosophical work that is displayed throughout the development of the work is as important to me as the hypothesis itself. Thus, even though the hypothesis could have had clearer contours, the lack of discussion of the hypothesis itself does not ultimately affect the main point of the thesis, namely the introduction of the Natural Problem and the typologies of answers to it.

Another reason for the absence of an in-depth discussion of objections to the hypothesis is that – arguably – the main objections against my hypothesis are actually objections against my initial assumptions (e.g., the existence/structure of "global quale", non-reductive physicalism). However, as I highlight throughout the work, I just conditionally assume these as a starting point, without pretending to establish or establish these assumptions more in solidly than required by the scope of the research work. It would be certainly valuable and very interesting to consider all these assumptions in turn, so as to further clarify and justify the basis of my hypothesis, but it would be unreasonable to pretend to satisfactorily answer all these complex questions all at once. My priority has been to sketch out the philosophical landscape arising out of a diachronic approach to consciousness intended as a biological phenomenon, and in order to do this I had to find a compromise between overall clarity and depth of analysis.

[3] The hypothesis holding that consciousness has a biological function does not answer the Hard Problem of how matter can develop phenomenal consciousness.

It is correct to say that my hypothesis to the Natural Problem does not provide an answer to the Hard Problem. I explicitly acknowledge several times in the thesis, conclusion included, that I am not directly concerned with the Hard Problem, meaning that I do not attempt to answer it. Indeed, that is why I distinguish the synchronic problems from the diachronic problems, by stating that I focus on the latter (section 1.5).

[4] There seems to be a lack of methodological reflection on what is the specific *philosophical* contribution to the hypothetical answer to the Natural Problem in terms of fitness enhancing, given that all the work is done with reference to the biological function of consciousness (i.e., by means of a *biological* explanation).

It is true that I invoke the biological function of consciousness in terms of fitness enhancing in order to try to answer the Natural Problem, and it is also true that this ultimately amounts to a biological explanation. In the light of the methodological naturalism that I assume (section 3.1), however, it is desirable to have this sort of ultimate explanation regarding biological phenomena. That is, since I argue that consciousness is actually a biological phenomenon, it is a good thing that the explanation be ultimately biological rather than philosophical. The specifically philosophical contribution to the explanation consists in showing how biological notions, such as the idea of biological function and biological explanations in general, could be applied to a non-standard object of biological enquiry and hotly debated topic in philosophy such as consciousness. More precisely, the specifically philosophical work lies in a careful presentation of the *explananda* (i.e., consciousness) as a special kind of biological phenomenon, namely as a biological phenomenon with an ontologically subjective mode of existence, and in the outlining of some obstacles that a successful biological explanation of consciousness has to overcome.

A first specific philosophical contribution to my hypothesis, thus, lies behind the explanation itself. An important part of the philosophical work consists in distinguishing the different strategies that one may pursue to explain what the biological function of consciousness might be, discussing and evaluating the potential for success of such explanatory strategies (i.e., bottom-up VS top-down). I do so extensively in sections 7.1 and 7.2 on the backdrop of the previously introduced assumptions and distinctions. This work is specifically philosophical because it does not consist in – say – questioning the empirical results of bottom-up research on the function of particular transitive mental states, but rather in questioning the general explanatory limits of any such research for the task at hand (i.e., to give an explanatory satisfactory answer to the Natural Problem).

A second specific philosophical contribution to my hypothesis (section 7.3) lies in the building of the theoretical scaffold that holds my naturalist hypothesis about the function of feeling together. This preparatory work is a philosophical rather than biological work because it consists in a theoretical step-by-step analysis of the constitutive elements required to give an explanatory answer to the Natural Problem.

More in detail, this includes:
a) The explicit distinction of the questions that need to be considered in order to construct a potentially explanatory hypothesis, namely:
   (i) What might be the biological causal role of being conscious *simpliciter*?
   (ii) Why is this biological causal role played by feeling rather than by something else?
b) The introduction, explanation, conceptual analysis and justification of terminological distinctions and notions in order to clarify and ease the understanding of the upcoming hypothesis. E.g., biological efficacy (section 7.3.1), algorithmic/heuristic strategy to maximize energy profit, token input stimulus, token global neural state, feeling tokens / phenomenally conscious states of mind, global quale (section 7.3.2), multiple realization, phenomenal similarities across global qualia (section 7.3.3), valence and valence spectrum (section 7.3.4).
c) The explicit distinction and discussion of two assumptions on which my hypothesis relies, namely:
   (i) Feeling beings can rely on a rough – but better than chance – pre-conceptual phenomenal individuation of situations that "feel good" or "feel bad" (section 7.3.5).
   (ii) On average, relying on valence feelings allows efficient behavior in virtually any situation (section 7.3.6).
d) The consideration of the philosophical problem of why the *ontologically subjective* mode of existence of valence feelings could be important to prompt biologically efficient behavior (section 7.3.7).

In conclusion, even though my tentative answer to the Natural Problem ultimately consists in a biological explanation, getting there requires and includes a great deal of philosophical work.

[5] By the introduction of the neologism "Global Quale" you introduce a substantial thesis, namely that some token global neural states are characterized by a global quale, and that creature consciousness cannot be reduced to individual conscious states. How do you support it? Even if we agree that (i) feeling exists, and (ii) intransitive creature consciousness has its own mental content, the argument for (iii) "the phenomenal content of intransitive creature consciousness can be individuated only by a global quale and not by many individual qualia" is not conclusive.

It is correct to affirm that postulating the existence of global quale is a substantial thesis. I support this thesis by arguing that since I do not assume a "state consciousness" approach, I do not consider the notion of qualia as applying to particular conscious mental states (section 4.3.4), and – pivotally – by means of a phenomenological argument saying that what philosophers usually define as particular individual qualia are unified in actual phenomenology (section 4.1.1). Probably the clearest exposition of my views on this matter lies at the end of section 1.3.2. Notice that I do not claim, as purportedly stated, that the phenomenal content of an intransitive globally conscious state of mind cannot be *individuated* or *described* (a posteriori) as being composed of several qualia. I only claim that our phenomenological experience at any one time is unified rather than scattered and fragmented, and that in this sense it is reasonable to consider one's overall conscious experience in such terms. As highlighted at the beginning of chapter 4, my endorsement of the terminological conventions I propose does not preclude that other conventions are less valuable. I just try to describe experience without blindly adopting the received vocabulary of philosophers interested in other issues than what I am interested in.

[6] It seems that you are talking about global qualia as of an atomic quale without internal structure. However you then claim that different feelings can share similar qualitative properties. If global qualia have no internal structure, however, feeling cannot have a qualitative structure of properties. So what is the right interpretation?

I do not consider global qualia as having no internal structure (or as being atomic). I claim that consciousness is phenomenally unified, that is, that a feeling appears phenomenally as unified rather than as being composed of several clearly distinct qualia. This however does not necessarily imply that it has an atomistic structure or no structure at all. Indeed, I claim that global qualia have qualitative properties, meaning that it is possible for the subject of experience to phenomenally identify similarities across qualitative properties of different feelings. One thing is how consciousness is present in phenomenal experience (as unified), another is how we can analyse such experience a posteriori. Toward the end of section 4.1.1 I say for example: "Consciousness presents itself phenomenologically as a much more complex, holistic and somewhat fuzzy property encompassing different sensory modalities and different sorts of mental states all at once". I do not see any contradiction between saying this and my later claims. Thus, I do not believe that my use of terminology is inconsistent throughout the work. I do not attempt to develop a theory about structure of experience. Nevertheless I suggest a possible line of investigation in terms of a mereological theory of conscious experience (footnote in section 7.3.3).

[7] Your rejection of the bottom-up approaches depends tacitly on the assumption of your substantial thesis. Dretske holds that:
(i) **Creature consciousness is composed by the phenomenal content of a being's conscious states.**
(ii) **The property of consciousness of these conscious states is an intrinsic property of these states (First-order theory of consciousness)**
(iii) **The phenomenal properties of these states on the contrary are not intrinsic, but relational properties (Representationalism)**
(iv) **The general function of phenomenally conscious states consists in making a being attending properties of objects.**

Your criticism of Dretske overlooks him in at least three respects. (1) For Dretske, at the level of Creature Consciousness and global quale, there is nothing to explain. Therefore one cannot reproach he should explain the function of creature consciousness without an argument and solely on the basis of terminological assumption. (2) Dretske is not interested in the function of visual or olfactory consciousness, but in the function of consciousness *tout court*, which consists in being able to perceive properties of certain objects in the environment. So Dretske provides a general theory of consciousness under an evolutionary perspective, not only related to subsystems. (3) Dretske represents an externalist type of consciousness theory according to which phenomenal properties of a conscious state are nothing but the represented properties of that conscious state. Since the representational content of states is individuated externally by Dretske (it is about nothing more than represented properties), *a forteriori* also the phenomenal character has to be individuated externalistically. This is a controversial position, it shows, however, that an internalism in terms of phenomenal consciousness cannot be accepted without further ado. You endorse however such internalism without arguments in the second step of the consideration (in terms of local supervenience thesis).

Regarding (1), I am not arguing that Dretske *should* explain the function of creature consciousness. I am reproaching him for underestimating the problem of determining the function of creature consciousness, claiming that such a task might not be as straightforward as he depicts it (see his quote at the beginning of section 7.2). I do so by suggesting that the results of a bottom-up analysis on the function of conscious states ought not be generalized and taken to suggest what the function of creature consciousness is without further argument. My criticism of Dretske is thus not aimed at showing that *he* should have done something else, but rather at suggesting that in order to answer the Natural Problem – which is a problem about the function of creature consciousness – it might not be sufficient to look at the function of conscious states. My ultimate goal, remember, is that of determining which methodological strategy (bottom-up or top-down) is more promising to determine what the biological function of feeling simpliciter might be.

Regarding (2), it is correct to affirm that Dretske is interested in the function of consciousness *tout court*. What I criticise Dretske for is not that he does not *attempt* or *aim* to provide a general theory of consciousness, but rather that his general theory about the function of being conscious of something is derived bottom-up from the evidence coming from a specific case of the lack of visual consciousness in blindsight, and this evidence – I claim – does not suffice to support his general theory. Thus, I do not claim that Dretske's *intentions* are mistaken, but rather that the argumentative strategy by which he tries to fulfill his intentions is not satisfactory. It is in this sense that I claim – contra Dretske – that the evidence coming from the analysis of blindsight alone may suffice to provide a theory of the function of consciousness for this specific subsystem, but nothing more (i.e., it does not alone suffice to justify Dretske's claim about the function of consciousness *tout court*).

Regarding (3), there are philosophers (e.g., Carruthers; Dennett; Dretske 1995; Lycan 1987,1996; Rosenthal; Tye 1995, 2000) who think that consciousness is explanatorily derived from intentionality. By these accounts, what consciousness has to do with intentionality depends on the prior general account of content or intentionality that one has, but there is no special issue regarding the internal or external fixation of the phenomenal character of experience, over and above what arises for mental content in general. By accepting that consciousness is explanatorily derived from intentionality and from some internationalizing account of consciousness one can try to claim (contra Chalmers, Levine, Nagel, etc.) that phenomenal consciousness does not pose a particular problem for reductive physicalist or materialist explanation. The main challenge of these philosophers consists in giving a natural scientific account of intentionality or mental representation, or, as Siewert puts it (2011, p. 52):

> The underlying thought is that a science of consciousness must adopt this strategy: first conceive of intentionality (or content or mental representation) in a way that separates it from consciousness, and see intentionality as the outcome of familiar (and non-intentional) natural causal processes. Then, by further specifying the kind of intentionality involved (in terms of its use, its sources, its content), we can account for consciousness. In other words: "naturalize" intentionality, then intentionalize consciousness, and mind has found its place in nature.

I announce that I do not consider the topics of intentionality and intentional mental content in the present work (footnotes at the beginning of section 2.1; section 4.1.4). One reason for this is that, contrary to the above-mentioned philosophers, I do not take consciousness (phenomenal character) to be derived from intentionality or mental content (see also Block 1990, 1995). I mention this explicitly in my short treatment of the Higher-Order theories of consciousness (section 4.1.5), arguing contra Rosenthal that I believe that we cannot separate the question of what it is for mental states to be conscious from the question of the sensory character of conscious mental states. Another reason for this is that the topic of the relation between consciousness and intentionality is far too wide and controversial to be satisfactorily treated in this work.

I grant that it would be a good idea to consider the problem of content externalism and internalism further, and that I could have argued more explicitly in favor of my assumption of internalism regarding phenomenal consciousness. My treatment of Dretske, however, is not directed against his starting point on these matters – namely the thesis that all conscious mental states have wide contents (phenomenal externalism) and that the phenomenal characters of these conscious states supervene on their contents (representationalism) – as much as against the claim that individuating the function of conscious vision (by the analysis of blindsight) suffices to individuate the function of consciousness *simpliciter*.

[8] Even though for good reasons you want to separate the metaphysical problem of consciousness from the Natural Problem, you do not succeed in this. The local supervenience thesis proceeds from an internalism in terms of consciousness and assumes the existence of Global Qualia. Apparently you do not succeed in keeping the Natural Problem free from substantial metaphysical assumptions. This is not particular surprising. It is especially problematic that the separation between these two problems leads you not to make explicit and to discuss the metaphysical assumptions in your treatment of the Natural Problem, but rather introduce them by suggesting you proceed "pragmatically". You distinguish the metaphysical and Natural Problem by mentioning the modal scope of such problems: the natural problem is a problem about the actual world, whereas the metaphysical deals with possible worlds. However in your argumentation (e.g., the end of section 7.3.3) you use the modal difference of actuality and possibility. This difference should however distinguish the metaphysical and the Natural Problem. Apparently the Natural Problem cannot be tackled without metaphysical assumptions.

Notice that I do not claim that I do not want to enter metaphysics at all. Neither do I say that the Natural Problem arises or can be answered independently from given metaphysical assumptions or argumentations. Indeed I explicitly say that that the boundaries between the stream of research "metaphysics", "mental content", and "consciousness" are often blurred and that the answer to one question depends at least partially on one's take on another question (section 2.1).

What I want to make clear from the outset is that the *goal* of the Natural Problem differs from the *goal* of the metaphysical problem. The metaphysical mind-body problem is aimed at the question of the relationship between the mental and the physical in general (across possible worlds). The Natural Problem, on the contrary, consists in finding a possible explanation to an actual contingent fact about nature, namely the current existence of consciousness, given some assumptions. It is in their *goal* that the two problems differ. Thus, my attempt to keep the two problems apart is motivated by an attempt to show that my goal is not to solve the metaphysical mind-body problem, but rather to introduce and unpack another one.

The reason why I try to highlight this is that objections directed towards the metaphysical conditional assumptions on the basis of which the Natural Problem arises (something that is conditionally assumed) are aiming at the wrong target. Strong objections against my hypothesis regarding the Natural Problem should show that, even granting my metaphysical assumptions, the argumentation is flawed and the conclusion does not follow. Thus, even if it is true that the Natural Problem arises only on the basis of the conditional acceptance of some substantial metaphysical assumptions (e.g., in my case, non-reductive physicalism, supervenience, the existence of global qualia, the presence of a mismatch between feeling types and neural state types), since I explicitly introduce these assumptions as my starting point, I do not see why this is problematic.

It is perhaps true that I could have made my metaphysical assumptions even more explicit, and it is true that I do not discuss them at length, but the reason for this is that my primary goal is not to prove that this or that metaphysical assumption is sound (metaphysical problem), but rather that on the basis of such assumptions there is an interesting philosophical problem arising (Natural Problem). It is fair to admit that in order to fully justify my position with respect to the Natural Problem one would have to carefully argue and defend the metaphysical assumptions, however this is an important job that cannot be carefully executed in a single piece of work.

My adoption of a pragmatic or practical attitude is explained by the fact that in order to tackle the Natural Problem we *have* to assume one of several possible metaphysical positions despite not knowing for sure which one correctly accounts for the actual world. Since taking a metaphysical stance is *required* to consider the Natural Problem, there are good practical reasons to assume a position that one takes to be more likely true than others (section 3.1), even without having a full metaphysical justification for such a position. I proceed pragmatically contra skeptics (section 3.2, conclusion of section 5.1) for the same reason: in order to tackle the Natural Problem one *has* to take a chance and endorse a position (e.g., on the basis of intuitions) even though one does not have any certainty that his metaphysical assumptions are correct. Pragmatism, in sum, is required to proceed towards the Natural Problem unless there is a widespread consensus on the metaphysical assumptions which, I believe, is not currently the case.

[9] Is non-reductive physicalism a consistent metaphysical position? How can physical "stuff" have properties that are ontologically distinct from, and irreducible to physical properties, and yet be *physical* stuff in contrast to something else? Does the neural substratum bear/instantiate two distinct types of properties (i.e., physical and non-physical properties)? Are non-physical properties ontologically distinct from the physical ones, although standing in a particular relationship to them (are you endorsing ontological dualism)?

In the present work I explicitly claim that I do not elaborate on non-reductive physicalism as a position in the metaphysical mind-body problem, and that I simply "endorse it conditionally as my metaphysical head quarter" (section 2.3):

> Given that my purpose is not to tackle the metaphysical mind-body problem, but rather to think about the Natural Problem of Consciousness I do not have to defend non-reductive physicalism *qua* metaphysical position. [...] I endorse non-reductive physicalism as being true for the sake of the argument and, given the conditional assumption of this metaphysical picture of reality, I proceed to introduce the Natural Problem of Consciousness.

Nevertheless, I think that non-reductive physicalism is a consistent and appealing position (see section 2.2.2 for an introduction). Physicalism is the view of reality being "fundamentally" formed by physical stuff. Depending on what we take "fundamentally" to mean, it seems reasonable to suggest that physical stuff (physical substance) could have different kinds of properties – e.g., physical and mental/phenomenal standing in a particular relation (e.g., of supervenience). More in detail:

(a) Physical stuff instantiating physical and non-physical properties is described as physical rather than mental by virtue of the fact that what makes physical stuff what it is, and not something else, are physical properties (i.e., physical properties are essential to physical stuff). Indeed, if we admitted that non-physical properties supervene on physical properties, in the absence

of physical properties there would be nothing left (i.e., there would be no non-physical stuff instantiating non-physical properties).
(b) I describe the two kinds of properties as ontologically distinct and non-physical properties as irreducible to physical properties because the two have a different modes of existence, even though the very existence of the subjective mode of existence (non-physical properties) depends on the existence of the objective mode of existence (physical properties), and not vice-versa. See section 2.1.3 for the claim that ontological distinctness does not require a full independence of mental properties on physical properties.
(c) Kim's mind-body dependence thesis holds that our psychological character is wholly dependent on and determined by our physical nature, and that what happens in our mental life is wholly dependent on, and determined by, what happens with our bodily processes. I take this to explicitly highlight a dependence relation between two things (namely, the mind depending on the body), while also affirming the ontological primacy, or priority, of the physical in relation to the mental. It is on the basis of the above thesis (and supervenience) that I conclude section 2.2.3 by explicitly saying that I assume "a form of non-reductive physicalism saying that mental states and events are wholly dependent on and determined by our physical (bodily) states and processes – reality being fundamentally formed by physical stuff -, and yet that mental properties are ontologically distinct from, and irreducible to physical properties". My position is therefore clearly and unambiguously, I believe, a form of ontological dualism about properties (and not a dualism about substances). I do not see why the dependence claim as stated above could ambiguously suggest the endorsement of a physicalist reductionism.

[10] Non-reductive physicalism is nowadays considered by many as plausible. You accept this thesis in your argumentation (section 2.3). However there is a question: if the original physicalism (by Lewis) was motivated as a direct solution to the problem of mental causation, given that the new non-reductive physicalism does not offer a direct solution of such problem, why should one still accept physicalism? The standard reply says that the natural sciences are based on physicalism. The motivation, however, is then circular: "I accept a form of physicalism because I accept a form of physicalism."

My motivation for assuming a form of physicalism (even if with the problem of mental causation lurking) is indeed that I want to propose a framework that is in principle compatible with natural science (section 2.1.2). The motivation for this choice is methodological: I hold that a philosophical proposal can be taken seriously only if (i) it starts off by acknowledging (or at least not contradicting) the principles that are legitimated by the natural science, or if (ii) it can reasonably prove those principles to be wrong (that is, by falsifying the existing theory). Since I do not intend nor try to falsify the current view offered by natural science, I assume its presuppositions as a starting point and try to make space for my suggestion within that framework. I tackle this point in my discussion of naturalism (section 3.1).

[11] It is crucial to your hypothesis that mental properties are causally efficacious for behavior, otherwise they could not enhance fitness. But you cannot stay neutral with respect to the issue of the causal completeness. Either (a) the Principle of Completeness holds (a version of Physicalism), and then the fitness of the being would be the same independently of whether phenomenal consciousness is causally efficacious (causal overdetermination) or whether it is epiphenomenal, or (b) the Principle of Completeness does not hold, phenomenal consciousness is causally efficacious alone, and so this is a version of ontological dualism.

First of all, an important cautionary note. The main goal of the dissertation, as said at the outset, is to "introduce and unpack" the question of why there are presently conscious beings at all (i.e., the Natural Problem). My thesis sets out not only one, but two possible answers to the Natural Problem, given the conditional assumption of non-reductive physicalism. Either feeling is epiphenomenal and feeling presently exists because it was a byproduct of something else, or feeling exists by virtue of the fact that it enhanced (and possibly still enhances) fitness (section 6.4). It is clearly stated that the reason why my hypothesis is based on the idea that feeling could enhance fitness is that, if this is possible, this would yield an explanatory answer to the Natural Problem. However, notice that in my conclusion I am careful not to overstate the merits of my hypothesis (section 7.3.8), and that in this sense the mapping of the philosophical terrain does not hinge directly on the metaphysical problem of mental causation (causal exclusion) and on the question of completeness.

This being said, it is fair to claim that the non-epiphenomenalist hypothesis is sound only if mental properties are somehow causally relevant for behavior, and this poses the question of how feeling could fit into the causation process. I explicitly claim that the presence of additional valence-based information makes an *indirect* causal difference by becoming part of one's token input stimulus thereby contributing to determining which token global state is going to ensue. I say that even if (positive) valence feelings do not contribute to guiding behavior by actively and directly *doing* something, they might nonetheless contribute to guiding behavior passively and indirectly, by being part of overall information on the basis of which action derives (section 7.3.6). This suggestion is supported by the assumption of causation as counterfactual dependence in chapter 8. Thus, I do not stay neutral to the topic of causal closure: I do endorse it.

The choice of options with respect to the issue of causal completeness, instead of being limited to options (a) or (b) could therefore be so completed:

(b) Causal completeness does not hold, feeling is alone causally efficacious. This would be a form of ontological dualism.

(a^1) Causal completeness holds and feeling is epiphenomenal. In this case we would have a form of non-reductive physicalism and a possible answer to the Natural Problem.

(a^2) Causal completeness holds and yet feeling is causally efficacious (overdetermination). In this case we would have a form of non-reductive physicalism in which an action can have both a complete and sufficient physical cause (a given neural state) and a complete and sufficient mental cause (feeling) that are equally causally relevant for the action in question. I do not endorse this option: I do not claim that feeling is a complete and sufficient cause for action, but only that actions can counterfactually depend on feelings.

(a^3) Causal completeness holds and feeling plays an indirect causal role. In this case we would have a form of non-reductive physicalism in which any action has a complete physical cause (a given neural state), but where *which* physical cause (neural state) is instantiated depends on the totality of contextual factors to which a being is exposed at any one moment (i.e., the token input stimulus). I claim that if there is a mismatch between feeling and neural types (section 7.3.3), how one feels (subjectively) in a given situation may add on to other (objective) factors and thereby contribute to determining which neural state is instantiated (and which not).

One may insist that I am only moving the worry of mental causation one step further (bringing us back to either a^1, a^2, or a^3). How can mental states cause a change in the token input stimulus? Against this, one could reply that the token input stimulus is not a physical state, but rather an informational state. If you cry while reading a book, there is a direct physical cause of the action of crying. That neural cause itself, however, would not have ensued in the absence of an appropriate supervening mental state (e.g., a feeling – say – of commotion or sadness), over and above the perception of physical inputs (the mere detection of visual signs on the paper). By means of learning and personal (ontogenetic) experience, the same objective input (the book) can acquire different meanings and our physical reaction to the same external objective stimuli can differ (see law of reinforcement, section 7.3.7).

I explicitly grant that my hypothesis is far from being uncontroversial, but at least it is philosophically engaged, and (if right) explanatorily richer than epiphenomenalism. Since the problem of mental causation is a hotly debated one, and no clear answer is in view, the problem is at any rate by no means unique to my approach.

[12] You stress that the ontological autonomy of the subjective perspective is for you a non-negotiable thesis. It follows from that, that there are subjective facts or properties. Feelings are such properties. Even if you argue that valence (as component of the Global Quale) makes an indirect causal difference and therefore can have a biological function, it remains unclear how a subjective and ontologically autonomous entity (Valence) could make a difference in an objective and ontologically autonomous process (input-condition and behavior). Of course, you would like to remain neutral with respect to the Hard Problem (somehow consciousness arises of matter, but how?). Instead you would like to answer the question "why consciousness emerged as biological function?" However the latter goes together with the possibility that consciousness can make a causal difference for behavior. On this central point one cannot only say that you do not know how that works, but that it works somehow. Otherwise this risks being as unsatisfactory an answer like Descartes' answer to Princess Elisabeth.

It is true that a fully satisfactory explanatory answer to the Natural Problem ought to say *how* consciousness can make a causal difference to behaviour. I do not claim to have given a fully convincing answer to this problem (section 7.3.8), and in this sense I bite the bullet and accept the criticism. Nevertheless I have, I believe, the merit of having tried to go at least some distance in formulating a hypothesis, and not to have claimed to have reached, or proved more than my hypothesis actually suggests.

What I have attempted to do is to suggest how, despite causal closure (i.e., physical action and behaviour having a complete physical cause: a global neural state), feeling could make an indirect causal difference to behaviour. I suggest that one's token global neural state is determined at any one time by one's token input stimulus (i.e., the totality of information one accesses at that time). Such information, I suggest, derives from the senses as well as, sometimes, from one's feelings. As my discussion of causal theories suggests (chapter 8), a counterfactual dependence of behaviour on the presence or absence of consciousness (in the presence of an asymmetry between feeling types and neural state types) might be sufficient to claim that consciousness plays a causal role.

I grant that my hypothesis in terms of how phenomenally-individuated information (valence) could make a difference to the input conditions is still at an embryonic stage. Indeed, I have to admit that during the development of the hypothesis (and still now) I found myself often stuck with this intuition that seems to be promising, but difficult to fully come to grips with. This may be because the hypothesis is wrong, or because I lack the knowledge and argumentative tools required to offer a solid argumentation with respect to this last (but important) step in the argument. I have to admit that this slippery end of the argument is quite frustrating, but on the other hand it is also incredibly stimulating – forcing me to push the research further ahead.

This being said, notice that even if my explanation falls short of giving or indicating a way to provide a fully satisfactory explanatory theory explaining the current presence of conscious beings in terms of fitness enhancement, this does not represent a major downfall. The main goal of the thesis was that of introducing and unpacking the Natural Problem. By identifying two viable kinds of answers (epiphenomenalism or consciousness having a causal function), I have therefore reached such a primary goal.

Bibliography

Allen, Grant (1881): "Sight and smell in vertebrates", *Mind* 6, No. 24, pp. 453–471.
Anscombe, Gertrude E.M. (1971): *Causality and Determination: An Inaugural Lecture*. Cambridge: Cambridge University Press.
Armstrong, David (1984): "Consciousness and causality", in D. Armstrong and N. Malcolm (eds.), *Consciousness and Causality*, Oxford: Blackwell, pp. 103–191.
Armstrong, David (1978): "What is consciousness?", *Proceedings of the Russellian Society* 3, pp. 65–76. Reprinted in *The Nature of the Mind and Other Essays*. Ithaca, NY: Cornell University Press (1981, pp. 55–67); and in N. Block et al. (eds.), *The Nature of Consciousness*, 1997, pp. 721–728.
Armstrong, David (1968): *A Materialist Theory of Mind*. London: Routledge and Kegan Paul.
Aydede, Murat (Spring 2013 Edition): "Pain", *The Stanford Encyclopedia of Philosophy*, Edward N. Zalta (ed.), URL =<http://plato.stanford.edu/archives/spr2013/entries/pain/>.
Baars, Bernard J. (1988): *A Cognitive Theory of Consciousness*. Cambridge: Cambridge University Press.
Barrett, Lisa (Feldman); Bliss-Moreau, Eliza (2009): "Affect as a psychological primitive", *Advances in Experimental Social Psychology* 41, pp. 167–218.
Barrett, Lisa (Feldman); Russell, James A. (1999): "Structure of current affect", *Current Directions in Psychological Science* 8, pp. 10–14.
Baumgartner, Michael (2008): "Regularity theories reassessed", *Philosophia* 36, No. 3, pp. 327–354.
Bayne, Tim (2007): "Conscious states and conscious creatures: explanations in the scientific study of consciousness", *Philosophical Perspectives* 21, pp. 1–22.
Beebee, Helen (2006): "Does anything hold the universe together?", *Synthese* 149, No. 3, pp. 509–533.
Beebee, Helen; Hitchcock, Christopher; Menzies, Peter (2009): *The Oxford Handbook of Causation*. Oxford: OUP.
Berkeley, George [1713]: *Three Dialogues Between Hylas and Philonous*. In R.S. Woolhouse (ed.), *Principles of Human Knowledge; and Three Dialogues Between Hylas and Philonous*, 1988, London: Penguin Books.
Berkeley, George [1710]: *Treatise Concerning the Principles of Human Knowledge*. In R.S. Woolhouse (ed.), *Principles of Human Knowledge; and Three Dialogues Between Hylas and Philonous*, 1988, London: Penguin Books.
Bickle, John (2003): *Philosophy and Neuroscience. A Ruthlessly Reductive Account*. Dordrecht: Kluwer.
Block, Ned (2011): "The higher order approach to consciousness is defunct", *Analysis* 71, pp. 419–431.
Block, Ned (1995): "On a confusion about a function of consciousness", *Behavioral and Brain Sciences* 18, pp. 227–287.
Block, Ned (1980a): "Troubles with functionalism". In N. Block (ed.), *Readings in the Philosophy of Psychology*, Volume 1, Cambridge, MA: Harvard University Press, pp. 268–305.
Block, Ned (1980b): "Are absent qualia impossible?", *Philosophical Review* 89, No. 2, pp. 257–274.
Block, Ned; Flanagan, Owen; Güzeldere, Güven (eds.)(1997): *The Nature of Consciousness*. Cambridge, MA: MIT Press.

Bogen, Joseph E. (1995): "On the neurophysiology of consciousness, part I: an overview", *Consciousness and Cognition* 4, pp. 52–62.
Boyd, Richard (1980): "Materialism without reductionism: what physicalism does not entail". In N. Block (ed.), *Readings in the Philosophy of Psychology*, vol. 1, Cambridge, MA: Harvard University Press, pp. 67–106.
Bornstein, Robert F.; Pittman, Thane S. (1992): *Perception Without Awareness*. New York: Guilford Press.
Brentano, Franz (1874): *Psychologie vom empirischen Standpunkt*. Leipzig: Meiner.
Bringsjord, Selmer; Noel, Ron (1998): "Why did evolution engineer consciousness?". In G. Mulhauser and J. Fetzer (eds.), *Evolving Consciousness*, 2000, Amsterdam: John Benjamins.
Brooks, Rodney (1991): "Intelligence without representation", *Artificial Intelligence* 47, pp. 139–159.
Burge, Tyler (1993): "Mind-body causation and explanatory practice". In J. Heil and A. Mele (eds.), *Mental Causation*, Oxford: Clarendon Press, pp. 97–120.
Byrne, Alex (2001a): "Intentionalism defended", *Philosophical Review* 110, pp. 199–240.
Byrne, Alex (2001b): "Review of Carruthers' phenomenal consciousness", *Mind* 110, pp. 1057–1062.
Byrne, Alex (1997): "Some like it HOT: consciousness and higher-order thoughts", *Philosophical Studies* 86, pp. 103–129.
Cacioppo, John T.; Gardner, Wendi L.; Berntson, Gary G. (1999): "The affect system has parallel and integrative processing components: form follows function", *Journal of Personality and Social Psychology* 76, pp. 839–855.
Campion, John; Latto, Richard; Smith, Yvonne M. (1983): "Is blindsight an effect of scattered light, spared cortex, and near threshold vision?", *Behavioral and Brain Science* 6, No. 3, pp. 423–486.
Carroll, James M.; Yik, Michelle S.M.; Russell, James A.; Barrett, Lisa (Feldman) (1999): "On the psychometric principles of affect", *Review of General Psychology* 3, pp. 14–22.
Carruthers, Peter (Fall 2011 Edition): "Higher-order theories of consciousness", *The Stanford Encyclopedia of Philosophy*, Edward N. Zalta (ed.), URL = <http://plato.stanford.edu/archives/fall2011/entries/consciousness-higher/>.
Carruthers, Peter (2005): *Consciousness: Essays From a Higher-Order Perspective*. Oxford: OUP.
Carruthers, Peter (2000): *Phenomenal Consciousness. A Naturalistic Theory*. Cambridge: Cambridge University Press.
Carruthers, Peter (1996): *Language, Thought and Consciousness*. Cambridge: Cambridge University Press.
Carruthers, Peter (1989): "Brute experience", *Journal of Philosophy* 86, pp. 258–269.
Chalmers, David (2016): "The combination problem for panpsychism". In G. Bruntrup and L. Jaskolla (eds.), *Panpsychism. Contemporary Perspectives*, 2016, Oxford: OUP.
Chalmers, David (2013): "Panpsychism and panprotopsychism", *Amherst Lecture in Philosophy 2013*. In T. Alter and Y. Nagasawa (eds.), *Consciousness in the Physical World: Essays on Russellian Monism*, 2013, Oxford: OUP, pp. 246–276.
Chalmers, David (ed.)(2002): *Philosophy of Mind. Classical and Contemporary Readings*. Oxford: OUP.
Chalmers, David (2000): "What is a neural correlate of consciousness". In T. Metzinger (ed.), *Neural Correlates of Consciousness*, Cambridge, MA: MIT Press, pp. 17–39.

Chalmers, David (1998): "On the search for the neural correlate of consciousness". In S. Hameroff, A. Kaszniak, and A.C. Scott (eds.), *Towards a Science of Consciousness*, Cambridge, MA: MIT Press, pp. 219–230.
Chalmers, David (1996): *The Conscious Mind: In Search of a Fundamental Theory*. New York: OUP.
Chalmers, David (1995): "Facing up to the problem of consciousness", *Journal of Consciousness Studies* 2, No. 3, pp. 200–219.
Charland, Louis (2005a): "Emotion experience and the indeterminacy of valence". In L.F. Barrett, P. Niedenthal, P. Winkielman (eds.), *Emotions and Consciousness*, New York: Guildford Press, pp. 231–254.
Charland, Louis (2005b): "The heat of emotion: valence and the demarcation problem", *Journal of Consciousness Studies* 12, No. 8–10, pp. 82–102.
Churchland, Patricia (1983): "Consciousness: the transmutation of a concept", *Pacific Philosophical Quarterly* 64, pp. 80–95.
Churchland, Patricia (1981): "Eliminative materialism and the propositional attitudes", *Journal of Philosophy* 78, pp. 67–90.
Cohen, Jonathan; Callender, Craig (2009): "A better best system account of lawhood", *Philosophical Studies* 145, pp. 1–34.
Collins, John; Hall, Ned; Paul, Laurie A. (2004): *Causation and Counterfactuals*. Cambridge, MA: MIT Press.
Colombetti, Giovanna (2005): "Appraising valence", *Journal of Consciousness Studies* 12, No. 8–10, pp. 103–126.
Coren, Stanley (2004): *How Dogs Think: Understanding the Canine Mind*. New York: Simon & Schuster, First Free Press.
Crick, Francis (1994): *The Astonishing Hypothesis: The Scientific Search for the Soul*. New York: Scribner's Sons.
Crick, Francis; Koch, Christof (1998): "Consciousness and neuroscience", *Cerebral Cortex* 375, pp. 121–123.
Crick, Francis; Koch, Christof (1995): "Are we aware of neural activity in primary visual cortex?", *Nature* 375, pp. 121–123.
Crick, Francis; Koch, Christof (1990): "Towards a neurobiological theory of consciousness", *Seminars in the Neurosciences* 2, pp. 263–275.
Damasio, Antonio (1999): *The Feeling of What Happens: Body and Emotion in the Making of Consciousness*. New York: Harcourt Brace & Co.
Damasio, Antonio (1994): *Descartes' Error: Emotion Reason and the Human Brain*. New York: Gossett/Putnam.
Davidson, Donald (1987): "Knowing one's own mind", *Proceedings and Addresses of the American Philosophical Association* 61, pp. 441–458.
Davies, Martin; Humphreys, Glyn W. (eds.)(1993): *Consciousness. Psychological and Philosophical Essays*. Oxford: Blackwell.
Dawkins, Richard (1986): *The Blind Watchmaker*. New York: W. Norton & Company.
Dawkins, Richard (1976): *The Selfish Gene*. Oxford: OUP.
Dehaene, Stanislas (2014): *Consciousness and the Brain. Deciphering How the Brain Codes Our Thoughts*. New York: Viking.
DeLancey, Craig (1996): "Emotion and the function of consciousness," *Journal of Consciousness Studies* 3, No. 5–6, pp. 492–499.

Dennett, Daniel C. (1992): "The self as the center of narrative gravity". In F. Kessel, P. Cole, and D.L. Johnson (eds.), *Self and Consciousness: Multiple Perspectives*, Hillsdale, NJ: Lawrence Erlbaum, pp. 103–115.
Dennett, Daniel C. (1991): *Consciousness Explained*. London: Allen Lane.
Dennett, Daniel C. (1990): "Quining qualia". In W. Lycan (ed.), *Mind and Cognition*, 1990, Oxford: Blackwell. pp. 519–548. Reprinted from A.J. Marcel and E. Bisiach (eds.), *Consciousness in Contemporary Science*, 1988, Oxford: OUP, pp. 42–77.
Dennett, Daniel C. (1988): "The evolution of consciousness". In J. Brockman (ed.), 1988, *The Reality Club*, Vol. iii, Prentice-Hall, pp. 3–99.
Dennett, Daniel C. (1987): "True believers". In D. Dennett, *The Intentional Stance*, 1987, Cambridge, MA: MIT Press, pp. 13–35.
Dennett, Daniel C. (1978a): "Toward a cognitive theory of consciousness". In C. Savage (ed.), *Perception and Cognition: Issues in the Foundations of Psychology*, Minneapolis: University of Minnesota Press. Reprinted in Dennett 1978b.
Dennett, Daniel C. (1978b): *Brainstorms*. Cambridge, MA: MIT Press.
Dennett, Daniel C. (1978c): "Why you can't make a computer that feels pain", *Synthese* 38, pp. 415–456.
Dennett, Daniel C. (1971): "Intentional systems", *Journal of Philosophy* 68, No. 4, pp. 87–106. Reprinted in Dennett 1978b.
Dennett, Daniel C.; Kinsbourne, Marcel (1992): "Time and the observer: the where and when of consciousness in the brain", *Behavioral and Brain Sciences* 15, pp. 187–247.
Descartes, René (1989): *Correspondance avec Elisabeth*. J.-M. Beyssade and M. Beyssade (eds.), Paris: Garnier-Flammarion.
Descartes, René [1644]: *Principia Philosophiae*. In E. Haldane and G. Ross (transl.), *The Principles of Psychology*, 1991, Cambridge: Cambridge University Press.
Descartes, René [1641]: *Meditationes de Prima Philosophia*. In J. Cottingham (transl.), *Meditations on First Philosophy*, 1996, Cambridge: Cambridge University Press.
Descartes, René [1637]: *Discours de la méthode*. In L. Lafleur (transl.), *Discourse on Method and Meditations*, 1960, Indianapolis, MI: Bobbs-Merrill Company.
Dowe, Phil (2000): *Physical Causation*. Cambridge: Cambridge University Press.
Dretske, Fred (1997): "What good is consciousness", *Canadian Journal of Philosophy* 27, No. 1, pp. 1–15.
Dretske, Fred (1995): *Naturalizing the Mind*. Cambridge, MA: MIT Press.
Dretske, Fred (1993): "Conscious experience", *Mind* 103, pp. 1–21.
Esfeld, Michael (2005): *La Philosophie de l'Esprit*. Paris: Armand Colin.
Feigl, Herbert (1958): "The "mental" and the "physical"", *Minnesota Studies in the Philosophy of Science* 2, pp. 370–497.
Feinberg, Todd; Mallatt, Jon (2016): *The Ancient Origins of Consciousness. How the Brain Created Experience*. Cambridge, MA: MIT Press.
Feyerabend, Paul (1963a): "Materialism and the mind-body problem", *Review of Metaphysics* 17, pp. 49–66.
Feyerabend, Paul (1963b): "Mental events and the brain", *Journal of Philosophy* 60, pp. 295–306.
Fisher, Ronald A. (1935): *The Design of Experiments*, London: Oliver & Boyd.
Flanagan, Owen (1992): *Consciousness Reconsidered*. Cambridge, MA: MIT Press.
Flanagan, Owen (1991): *The Science of the Mind*, 2nd ed., Cambridge, MA: MIT Press.

Flanagan, Owen; Polger, Thomas (1995): "Zombies and the function of consciousness", *Journal of Consciousness Studies* 2, pp. 313–321.
Flohr, Hans (1995): "Sensations and brain processes", *Behavioral Brain Research* 71, pp. 157–161.
Fodor, Jerry (1974): "Special sciences", *Synthese* 28, pp. 77–115.
Fodor, Jerry (1968): *Psychological Explanation*. New York, NY: Random House.
Foster, John (1991): *The Immaterial Self. A Defence of the Cartesian Dualist Conception of the Mind*. London: Routledge.
Foster, John (1989): "A defense of dualism". In J. Smythies and J. Beloff (eds.), *The Case for Dualism*, 1989, Charlottesville, VA: University of Virginia Press, pp. 1–23.
Freud, Sigmund [1933]: *New Introductory Lectures on Psychoanalysis*. In J. Strachey (transl.), *Complete Psychological Works of Sigmund Freud*, Vol.20, 1964, London: Hogarth Press.
Gardner, Howard (1985): *The Mind's New Science*. New York: Basic Books.
Gassendi, Pierre [1658]: *Opera Omnia*. Six Volumes. Reproduction of 1658 edition with introduction by Tullio Gregory, Lyons: Sumptibus Laurentii Anisson & Ioan. Bapt. Devenet; Reproduced in 1964, Stuttgart-Bad Cannstatt: Friedrich Frommann Verlag.
Gennaro, Rocco (ed.) (2004): *Higher-Order Theories of Consciousness*. Amsterdam and Philadelphia: John Benjamins.
Gennaro, Rocco (1995): *Consciousness and Self-consciousness: A Defense of the Higher-Order Thought Theory of Consciousness*. Amsterdam and Philadelphia: John Benjamins.
Gould, Stephen J. (1997): "The exaptive excellence of spandrels as a term and prototype", *Proceedings of the National Academy of Sciences USA* 94, pp. 10750–10755.
Gould, Stephen J.; Lewontin, Richard (1979): "The spandrels of San Marco and the panglossian paradigm: a critique of the adaptationist programme", *Proceedings of the Royal Society of London,* Series B 205, pp. 581–598.
Gould, Stephen J.; Vrba, Elisabeth (1982): "Exaptation: a missing term in the science of form", *Paleobiology* 8, No. 1, pp. 4–15.
Güzeldere, Güven (1997): "The many faces of consciousness: a field guide". In N. Block et al. (eds.), *The Nature of Consciousness*, Cambridge, MA: MIT press, pp. 1–67.
Harnad, Steven (2000): "Turing indistinguishability and the blind watchmaker". In G. Mulhauser and J. Fetzer (eds.), *Evolving Consciousness*, 2000, Amsterdam: John Benjamins.
Hasker, William (1999): *The Emergent Self*. Ithaca, NY: Cornell University Press.
Heidegger, Martin [1927]: *Sein und Zeit*. In J. Macquarrie and E. Robinson (transl.), *Being and Time,* 1962, New York: Harper and Row.
Hellman, Geoffrey P.; Thompson, Frank W. (1975): "Physicalism: ontology, determination and reduction", *Journal of Philosophy* 72, pp. 551–564.
Hill, Christopher S.; McLaughlin, Brian P. (1998): "There are fewer things in reality than are dreamt of in Chalmers' philosophy", *Philosophy and Phenomenological Research* 59, No. 2, pp. 445–454.
Hobbes, Thomas [1655]: *De Corpore*. Chapters 1–6. In A.P. Martinich (transl.), *Part I of De Corpore*, 1981, New York: Abaris Books.
Hobbes, Thomas [1651]: *Leviathan*. In E. Curley (ed.), *Leviathan, with selected variants from the Latin edition of 1668*, 1994, Indianapolis: Hackett.
Hobson, John Allan (1997): "Consciousness as state-dependent phenomenon." In J. Cohen and J. Schooler (eds.), *Scientific Approaches to Consciousness*. Mahwah, NJ: Lawrence Erlbaum, pp. 379–396.

Hume, David (1888)[1739]: *A Treatise of Human Nature*. Edited by L.A. Selby-Bigge. Oxford: Clarendon Press.
Hume, David (2008)[1748]: *An Enquiry concerning Human Understanding*. Edited by Peter Millican. Oxford: Oxford World's Classics.
Humphrey, Nicholas (1992): *A History of the Mind: Evolution and the Birth of Consciousness*. New York: Simon and Schuster.
Humphrey, Nicholas (1974): "Vision in a monkey without striate cortex: a case study", *Perception* 3, pp. 241–255.
Humphrey, Nicholas (1972): "Seeing and nothingness", *New Scientist* 53, pp. 682–684.
Humphrey, Nicholas (1970): "What the frog's eye tells the monkey's brain", *Brain, Behaviour and Evolution* 3, pp. 324–337.
Husserl, Edmund [1913]: *Ideen au einer reinen Phänomenologie und phänomenologischen Philosophie*. In W. Boyce Gibson (transl.), *Ideas: General Introduction to Pure Phenomenology*, 1931, New York: MacMillan.
Huxley, Thomas (1866): *Lessons on Elementary Physiology* 8. London: MacMillan & Co.
Hyslop, Alec (2016): "Other Minds", *The Stanford Encyclopedia of Philosophy* (spring 2016 edition), Edward N. Zalta (ed.), URL = <https://plato.stanford.edu/archives/spr2016/entries/other-minds/>.
Jackson, Frank (2004): "Why we need A-intensions", *Philosophical Studies* 118, pp. 257–277.
Jackson, Frank (1994): "Finding the mind in the natural world". In R. Casati, B. Smith, and G. White (eds.), *Philosophy and the Cognitive Sciences: Proceedings of the 16th International Wittgenstein Symposium*, Vienna: Verlag Holder-Pinchler-Tempsky, pp. 101–112. Reprinted in N. Block et al. (eds.), *The Nature of Consciousness*, 1997, Cambridge, MA: MIT Press, pp. 483–491.
Jackson, Frank (1982): "Epiphenomenal qualia", *Philosophical Quarterly* 32, No. 127, pp. 127–136.
James, William [1890]: *The Principles of Psychology*. Reprint edition, Vol.1 and 2, 2010, New York: Digireads.com Publishing.
James, William (1884): "What is an emotion?", *Mind* 9, pp. 188–205.
Kant, Immanuel [1787]: *Kritik der reinen Vernunft*. In N. Kemp Smith (transl.), *Critique of Pure Reason*, 1929, New York: MacMillan.
Kim, Jaegwon (2011): *Philosophy of Mind*. Boulder, CO: Westview Press.
Kim, Jaegwon (2007): "Causation and mental causation". In B. McLaughlin and J. Cohen (eds.), *Contemporary Debates in Philosophy of Mind*, 2007, Oxford: Blackwell Publishing, pp. 228–242.
Kim, Jaegwon (2005): *Physicalism or Something Near Enough*. Princeton: Princeton University Press.
Kim, Jaegwon (2003): "Blocking causal drainage and other maintenance chores with mental causation", *Philosophy and Phenomenological Research* 67, No. 1, pp. 151–176.
Kim, Jaegwon (1998): *Mind in a Physical World*. Cambridge, MA: MIT Press.
Kim, Jaegwon (ed.)(1993): *Supervenience and Mind: Selected Philosophical Essays*. Cambridge: Cambridge University Press.
Kim, Jaegwon (1989): "The myth of non-reductive physicalism", *Proceedings and Addresses of the American Philosophical Association* 63, No. 3, pp. 31–47. Reprinted in Kim 1993.
Kimura, Motoo (1983): *The Neutral Theory of Molecular Evolution*. Cambridge: Cambridge University Press.

Kirk, Robert (2015): "Zombies", *The Stanford Encyclopedia of Philosophy* (summer 2015 edition), Edward N. Zalta (ed.), URL = <https://plato.stanford.edu/archives/sum2015/entries/zombies/>.
Kistler, Max (2006): *Causation and Laws of Nature*. New York: Routledge.
Koch, Christof (2012): *Consciousness: Confessions of a Romantic Reductionist*. Cambridge, MA: MIT Press.
Koch, Christof (2004): *The Quest of Consciousness: A Neurobiological Approach*. Englewood, CO: Roberts & Company Publishers.
Kriegel, Uriah (2003): "Consciousness: phenomenal consciousness, access consciousness, and scientific practice". In P. Thagard (ed.), *Handbook of Philosophy of Psychology and Cognitive Science*, 2006, Amsterdam: North-Holland, pp. 195–217.
Kripke, Saul (1971): "Naming and necessity". In D. Davidson and G. Harman (eds.), *Semantics of Natural Language*, 1971, Dodrecht: Reidel, pp. 253–355.
Kripke, Saul (1980): *Naming and Necessity*. Cambridge, MA: Harvard University Press.
Larsen, Randy J.; Diener, Edward (1992): "Promises and problems with the circumplex model of emotion". In M. Clark (ed.), *Emotion*, 1992, Newbury Park, Ca: Sage Publications, pp. 25–29.
Larsen, Jeff T.; McGraw, A. Peter; Cacioppo, John T. (2001): "Can people feel happy and sad at the same time?", *Journal of Personality and Social Psychology* 81, pp. 684–696.
Larsen, Jeff T.; McGraw, A. Peter; Mellers, Barbara A.; Cacioppo, John T. (2004): "The agony of victory and thrill of defeat: mixed emotional reactions to disappointing wins and relieving losses", *Psychological Science* 15, pp. 325–330.
Laureys, Steven; Piret, Sonia; Ledoux, Didier (2005): "Quantifying consciousness", *Lancet Neurology* 4, No. 12, pp. 789–90.
Laureys, Steven; Majerus, Steve; Moonen, Gustave (2002): "Assessing consciousness in critically ill patients". In J. Vincent (ed.), *Yearbook of Intensive Care and Emergency Medicine*, 2002, Berlin/Heidelberg: Springer-Verlag, pp. 715–727.
Leibniz, Gottfried W. [1686]: *Discours de métaphysique*. In D. Garter and R. Aries (transl.), *Discourse on Metaphysics*, 1991, Indianapolis: Hackett.
Leibniz, Gottfried W. [1714]: *La Monadologie*. In R. Lotte (transl.), *The Monadology*, 1925, Oxford: OUP.
Lewis, David K. (1986): *On the Plurality of Worlds*. Oxford: Blackwell.
Lewis, David K. (1966): "An argument for the identity theory", *Journal of Philosophy* 63, pp. 17–25.
Lewontin, Richard (1970): "The units of selection", *Annual Review of Ecology and Systematics* 1, pp. 1–18.
Levine, Joseph (2001): *Purple Haze: The Puzzle of Consciousness*. Oxford/New York: OUP.
Levine, Joseph (1993): "On leaving out what it's like". In M. Davies and G. Humphreys (eds.), *Consciousness. Psychological and Philosophical Essays*, 1993, Oxford: Blackwell, pp. 121–136. Reprinted in N. Block et al. (eds.), *The Nature of Consciousness*, 1997, Cambridge, MA: MIT Press, pp. 543–555.
Levine, Joseph (1983): "Materialism and qualia: the explanatory gap", *Pacific Philosophical Quarterly* 64, pp. 354–361.
Lindquist, Kristen A.; Satpute, Ajay B.; Wager, Tor D.; Weber, Jochen; Barrett, Lisa (Feldman) (2016): "The brain basis of positive and negative affect: evidence from a meta-analysis of the human neuroimaging literature", *Cerebral Cortex* 26(5), pp. 1910–22.

Llinás, Rodolfo R.; Ribary, Urs; Joliot, Marc; Wang, Xiao-Jing (1994): "Content and context in temporal thalamocortical binding". In G. Buzaki, R. Llinás, W. Singer, A. Berthoz, and Y. Christen (eds.), *Temporal Coding in the Brain*, 1994, Berlin: Springer-Verlag, pp. 251–272.
Locke, John [1688]: *An Essay on Human Understanding*. In A.C. Fraser (ed.), *An Essay on Human Understanding*, 1959, New York: Dover Publishing.
Loewer, Barry (2007a): "Counterfactuals and the second law". In H. Price and R. Corry (eds.), *Causation, Physics, and the Constitution of Reality: Russell's Republic Revisited*, 2007, Oxford: OUP, pp. 293–326.
Loewer, Barry (2007b): "Mental causation, or something near enough". In B. McLaughlin and J. Cohen (eds.), *Contemporary Debates in Philosophy of Mind*, 2007, Oxford: Blackwell Publishing, pp. 243–264.
Loewer, Barry (2002): "Comments on Jaegwon Kim's Mind in a Physical World", *Philosophy and Phenomenological Research* 65, pp. 655–663.
Loewer, Barry (2001): "From physics to physicalism". In *Physicalism and its Discontents*, 2001, Cambridge: Cambridge University Press, pp. 37–56.
Lowe, Jonathan (2003): "Physical causal closure and the invisibility of mental causation". In S. Walter and H.-D. Heckmann (eds.), *Physicalism and Mental Causation*, 2003, Exeter: Imprint Academic, pp. 137–154.
Lowe, Jonathan (2000): "Causal closure principles and emergentism," *Philosophy* 75, pp. 571–585.
Lycan, William (2004): "The superiority of HOP to HOT". In R. Gennaro (ed.), *Higher-Order Theories of Consciousness,* 2004, Amsterdam and Philadelphia: John Benjamins, pp. 93–114.
Lycan, William (1996): *Consciousness and experience*. Cambridge, MA: MIT Press.
Lycan, William (1990): "Consciousness as internal monitoring", *Philosophical Perspectives* 9, pp. 1–14.
Lycan, William (1987): *Consciousness*. Cambridge, MA: MIT Press.
Lycan, William; Pappas, George (1972): "What is eliminative materialism?", *Australasian Journal of Philosophy* 50, pp. 149–159.
Mancuso, Stefano; Viola, Alessandra (2013): *Verde Brillante: Sensibilità e intelligenza del mondo vegetale*. Firenze: Giunti Editore.
Marcel, Anthony J. (1988): "Phenomenal experience and functionalism". In A.J. Marcel and E. Bisiach (eds.), 1988, *Consciousness in Contemporary Science*. Oxford: OUP, pp. 121–158.
Marcel, Anthony J. (1986): "Consciousness and processing: choosing and testing a null hypothesis", *Behavioral and Brain Sciences* 9, pp. 40–44.
Marcel, Anthony J.; Bisiach, Edoardo (eds.)(1988): *Consciousness in Contemporary Science*. New York: OUP.
McGinn, Colin (1991): *The Problem of Consciousness*. Oxford: Blackwell.
McGinn, Colin (1989): "Can we solve the mind-body problem?", *Mind* 98, pp. 349–366.
McLaughlin, Brian P.; Bennett, Karen (spring 2014 edition): "Supervenience", *The Stanford Encyclopedia of Philosophy* , Edward N. Zalta (ed.), URL = <http://plato.stanford.edu/archives/spr2014/entries/supervenience/>.
Mellor, David (1995): *The Facts of Causation*. London: Routledge.
Mellor, David (1971): *The Matter of Chance*. Cambridge: Cambridge University Press.
Merker, Bjorn (2007): "Consciousness without a cerebral cortex: a challenge for neuroscience and medicine", *Behavioural and Brain Sciences* 30, pp. 63–81.

Merleau-Ponty, Maurice [1945]: *Phénoménologie de la Perception*. In Colin Smith (transl.), *Phenomenology of Perception*, 1962, London: Routledge and Kegan Paul.
Metzinger, Thomas (ed.)(2000): *Neural Correlates of Consciousness*. Cambridge, MA: MIT Press.
Mill, James (1869)[1829]: *Analysis of the Phenomena of the Human Mind*. London: Longman, Green, Reader & Dyer.
Mill, John Stuart (1963)[1843]: *A System of Logic, Ratiocinative and Inductive*. In J.M. Robson (ed.), *Collected Works of John Stuart Mill*, volume 7–8, Toronto: University of Toronto Press.
Mill, John Stuart (1963)[1865]: *An Examination of Sir William Hamilton's Philosophy*. In J.M. Robson (ed.), *Collected Works of John Stuart Mill*, vol. 9, Toronto: University of Toronto Press.
Mills, Eugene (1996): "Interactionism and overdetermination", *American Philosophical Quarterly* 33, No. 1, pp. 105–117.
Milner, A. David (1995): "Cerebral correlates of visual awareness", *Neuropsychologia* 33, pp. 1117–1130.
Milner, A. David; Goodale, Melvyn A. (1995): *The Visual Brain in Action*. Oxford: OUP.
Milner, A. David; Rugg, Michael D. (eds.)(1992): *The Neuropsychology of Consciousness*. London: Academic Press.
Mulhauser, Gregory R.; Fetzer, James H. (eds.)(2000): *Evolving Consciousness*. Amsterdam: John Benjamins.
Mumford, Stephen; Anjum, Rani Lill (2013): *Causation: a Very Short Introduction*. Oxford: OUP.
Nagel, Thomas (1986): *The View from Nowhere*. Oxford: OUP.
Nagel, Thomas (1979): "Panpsychism". In *Mortal Questions*, 1979, Cambridge: Cambridge University Press.
Nagel, Thomas (1974): "What is it like to be a bat?", *Philosophical Review* 83, No. 4, pp. 435–450.
Neander, Karen (1998): "The division of phenomenal labor: a problem for representational theories of consciousness", *Philosophical Perspectives* 12, pp. 411–434.
Neisser, Ulric (1965): *Cognitive Psychology*. Englewood Cliffs: Prentice Hall.
Nelkin, Norton (1996): *Consciousness and the Origins of Thought*. Cambridge: Cambridge University Press.
Nelkin, Norton (1989): "Unconscious sensations", *Philosophical Psychology* 2, pp. 129–141.
Newman, James; Baars, Bernard J. (1993): "A neural attentional model of access to consciousness: a global workspace perspective", *Concepts in Neuroscience* 4, pp. 255–290.
Nichols, Shaun; Grantham, Todd (2000): "Adaptive complexity and phenomenal consciousness", *Philosophy of Science* 67, pp. 648–670.
Norris, Catherine J.; Gollan, Jackie; Berntson, Gary G.; Cacioppo, John T. (2010): "The current status of research on the structure of evaluative space", *Biological Psychology* 84, pp. 422–436.
Papineau, David (Spring 2009 Edition): "Naturalism", *The Stanford Encyclopedia of Philosophy*, Edward N. Zalta (ed.), URL = <http://plato.stanford.edu/archives/spr2009/entries/naturalism/>.
Papineau, David (2002): *Thinking About Consciousness*. Oxford: OUP.
Papineau, David (1995): "The antipathetic fallacy and the boundaries of consciousness". In T. Metzinger (ed.), *Conscious Experience*, 1995, Exeter: Imprint Academic, pp. 259–270.
Penfield, Wilder (1975): *The Mystery of the Mind: A Critical Study of Consciousness and the Human Brain*. Princeton: Princeton University Press.

Penrose, Roger (1994): *Shadows of the Mind*. Oxford: OUP.
Penrose, Roger (1989): *The Emperor's New Mind: Computers, Minds and the Laws of Physics*. Oxford: OUP.
Pinker, Steven (1997): *How the Mind Works*. New York: W.W. Norton & Company.
Place, Ullin T. (1956): "Is consciousness a brain process?", *British Journal of Psychology* 47, No. 1, pp. 44–50.
Pöppel, Ernst; Held, Richard; Frost, Douglas (1973): "Residual visual function after brain wounds involving the central visual pathways in man", *Nature* 243, pp. 295–296.
Popper, Karl; Eccles, John (1977): *The Self and Its Brain*. Berlin: Springer.
Prinz, Jesse (2005): "Are emotions feelings?", *Journal of Consciousness Studies* 12, No. 8–10, pp. 9–25.
Prinz, Jesse (2004a): "Which emotions are basic?". In D. Evans and P. Cruse (eds.), *Emotion, Evolution, and Rationality*, 2004, Oxford: OUP, pp. 69–87.
Prinz, Jesse (2004b): *Gut Reactions*. New York: OUP.
Putnam, Hilary (1973): "Philosophy and our mental life". Reprinted in *Mind Language and Reality: Philosophical Papers*, 1975, Cambridge: Cambridge University Press, pp. 291–303.
Putnam, Hilary (1967): "The nature of mental states". Reprinted in *Mind Language and Reality: Philosophical Papers*, 1975, Cambridge: Cambridge University Press, pp. 429–440.
Ramachandran, Vilayanur; Hirnstein, William (1998): "Three laws of qualia: what neurology tells us about the biological functions of consciousness", *Journal of Consciousness Studies* 4, No. 4–5, pp. 429–457.
Ratcliffe, Matthew (2005): "The feeling of being", *Journal of Consciousness Studies* 12, No. 8–10, pp. 43–60.
Rorty, Richard (1965): "Mind-body identity, privacy, and categories", *Review of Metaphysics* 9, pp. 24–54.
Rosenberg, Gregg (2004): *A Place for Consciousness: Probing the Deep Structure of the Natural World*. New York: OUP.
Rosenthal, David (2011): "Exaggerated reports: reply to Block", *Analysis* 71, No. 3. pp. 431–437.
Rosenthal, David (2005): *Consciousness and Mind*. New York: OUP.
Rosenthal, David (1993): "Thinking that one thinks". In M. Davies and G. Humphreys (eds.), *Consciousness: Psychological and Philosophical Essays*, 1993, Oxford: Blackwell, pp. 196–223.
Rosenthal, David (1991): "The independence of consciousness and sensory quality", *Philosophical Issues* 1, pp. 15–36. Reprinted in E. Villanueva (ed.), *Consciousness*, 1991, Atascadero, CA: Ridgeview Publishing.
Rosenthal, David (1990): "A theory of consciousness", Report No. 40, Research Group on Mind and Brain, ZiF, University of Bielefeld. Reprinted in N. Block et al. (eds.), *The Nature of Consciousness*, 1997, Cambridge, MA: MIT Press, pp. 729–753.
Rosenthal, David (1986): "Two concepts of consciousness", *Philosophical Studies* 49, pp. 329–359.
Russell, Bertrand (1959): *My Philosophical Development*. London: George Allen & Unwin.
Russell, Bertrand (1956a): "Mind and matter." In Russell 1956b, pp. 135–153.
Russell, Bertrand (1956b): *Portraits from Memory and Other Essays*. London: George Allen & Unwin.
Russell, Bertrand (1927a): *The Analysis of Matter*. London: Kegan Paul.
Russell, Bertrand (1927b): *An Outline of Philosophy*. London: George Allen & Unwin.
Russell, Bertrand (1921): *The Analysis of Mind*. London: George Allen & Unwin.

Russell, Bertrand (1919): "On propositions: what they are and how they mean", *Proceedings of the Aristotelian Society*, Supplementary Volumes, Vol. 2, pp. 1–43.

Russell, Bertrand (1913): "On the notion of cause", *Proceedings of the Aristotelian Society* 13, pp. 1–26. Reprinted in S. Mumford (ed.), *Russell on Metaphysics*, 2003, London: Routledge, pp. 164–182.

Russell, James A. (2005): "Emotion in human consciousness is built on core affect", *Journal of Consciousness Studies* 12, No. 8–10, pp. 26–42.

Russo, Andrew (2013): *A Defense of Nonreductive Mental Causation*. Dissertation, The University of Oklahoma.

Ryle, Gilbert (1949): *The Concept of Mind*. London: Hutchinson.

Seager, William; Allen-Hermanson, Sean (Fall 2013 Edition): "Panpsychism", *The Stanford Encyclopedia of Philosophy*, Edward N. Zalta (ed.), URL = <http://plato.stanford.edu/archives/fall2013/entries/panpsychism/>.

Searle, John R. (2004): *Mind. A Brief Introduction*. New York: OUP.

Searle, John R. (1997): *The Mystery of Consciousness*. New York: NYREV.

Searle, John R. (1992): *The Rediscovery of the Mind*. Cambridge, MA: MIT Press.

Searle, John R. (1990): "Who is computing with the brain?", *Behavioral and Brain Sciences* 13, No. 4, pp. 632–642.

Sheinberg, David L.; Logothetis, Nikos K. (1997): "The role of temporal cortical areas in perceptual organization", *Proceedings of the National Academy of Sciences* 94, pp. 3408–3413.

Shoemaker, Sydney (1990): "Qualities and qualia: what's in the mind", *Philosophy and Phenomenological Research* 50, pp. 109–131.

Shoemaker, Sydney (1981): "Absent qualia are impossible", *Philosophical Review* 90, pp. 581–599.

Shoemaker, Sydney (1975): "Functionalism and qualia", *Philosophical Studies* 27, pp. 291–315.

Siewert, Charles (Fall 2011 Edition): "Consciousness and Intentionality", *The Stanford Encyclopedia of Philosophy*, Edward N. Zalta (ed.), URL=<http://plato.stanford.edu/archives/fall2011/entries/consciousness-intentionality/>.

Skinner, B. Frederic (1953): *Science and Human Behavior*. New York: MacMillan.

Smart, John J.C. (1959): "Sensations and brain processes", *Philosophical Review* 68, pp. 141–156.

Sober, Elliott (2000): *Philosophy of Biology*. 2nd Edition. Boulder, CO: Westview Press.

Sosa, Ernest; Tooley, Michael (1993): *Causation*. Oxford: OUP.

Spinoza, Baruch [1677]: *Ethics*. In E. Curley (transl.), *The Collected Writings of Spinoza*, volume 1, 1985, Princeton: Princeton University Press.

Stich, Steven (1978): "Beliefs and sub-doxastic states", *Philosophy of Science* 45, pp. 499–458.

Stout, George (1899): *Manual of Psychology*. New York: Hinds, Noble, and Eldredge Publishers.

Strawson, Galen (1994): *Mental Reality*. Cambridge, MA: MIT Press.

Suárez, Mauricio (2011): *Probabilities, Causes and Propensities in Physics*. Dordrecht: Springer.

Sutherland, Stuart (1995): "Consciousness". In *MacMillan Dictionary of Psychology*, Second edition, London: The MacMillan Press.

Swinburne, Richard (1986): *The Evolution of the Soul*. Oxford: OUP.

Teasdale, Graham; Jennett, Bryan (1974): "Assessment of coma and impaired consciousness. A practical scale", *Lancet* 2, pp. 81–84.

Thorndike, Edward (1927): "The law of effect", *The American Journal of Psychology* 39, No. 1/4, pp. 212–222.

Thorndike, Edward (1911): *Animal Intelligence: Experimental Studies*. Lewiston, NY: Macmillan Press.
Titchener, Edward (1915): *A Beginner's Psychology*. New York: Macmillan.
Titchener, Edward (1901): *An Outline of Psychology*. New York: Macmillan.
Tononi, Giulio (2008): "Consciousness as integrated information: a provisional manifesto", *Biological Bulletin* 215, No. 3, pp. 216–242.
Trumbull Ladd, George (1909): *Psychology: Descriptive and Explanatory*. New York: C. Scribner's Sons.
Tye, Michael (2000): *Consciousness, Color, and Content*. Cambridge MA: MIT Press.
Tye, Michael (1995): *Ten Problems of Consciousness*. Cambridge, MA: MIT Press.
Van Gulick, Robert (Spring 2014 Edition): "Consciousness", *The Stanford Encyclopedia of Philosophy*, Edward N. Zalta (ed.), URL = <http://plato.stanford.edu/archives/spr2014/entries/consciousness/>.
Van Gulick, Robert (2004): "Higher-order global states HOGS: an alternative higher-order model of consciousness." In R. Gennaro (ed.), *Higher-Order Theories of Consciousness*, 2004, Amsterdam and Philadelphia: John Benjamins, pp. 67–92.
Van Gulick, Robert (2000): "Inward and upward: reflection, introspection and self-awareness", *Philosophical Topics* 28, pp. 275–305.
Van Gulick, Robert (1989): "What difference does consciousness make?", *Philosophical Topics* 17, pp. 211–223.
Van Gulick, Robert (1985): "Physicalism and the subjectivity of the mental", *Philosophical Topics* 13. pp. 51–70.
Velmans, Max (2007): "The co-evolution of matter and consciousness", *Synthesis Philosophica* 22, pp. 273–282.
von Helmholtz, Hermann [1910]: *Handbuch der physiologischen Optik* (3rd Ed.). In J. Southall (transl.), *Treatise on Physiological Optics*, 3 volumes, 1924, New York: Optical Society of America.
Watson, John (1924): *Behaviorism*. New York: W.W. Norton.
Watson, David; Tellegen, Auke (1985): "Toward a consensual structure of mood", *Psychological Bulletin* 98, pp. 219–235.
Weiskrantz, Lawrence (1997): *Consciousness Lost and Found*. Oxford: OUP.
Weiskrantz, Lawrence (1986): *Blindsight. A Case Study and Implications*. Oxford: OUP.
Wilberg, Jonah (2010): "Consciousness and false HOTs", *Philosophical Psychology* 23, pp. 617–638.
Wild, Markus (2008): *Tierphilosophie zur Einführung*. Hamburg: Junius Verlag.
Wilkes, Kathleen V. (1988): "Yishi, duo, us and consciousness". In A. Marcel and E. Bisiach (eds.), *Consciousness in Contemporary Science*, 1988, New York: OUP.
Wilkes, Kathleen V. (1984): "Is consciousness important?", *British Journal for the Philosophy of Science* 35, pp. 223–243.
Wundt, Wilhelm (1897): *Outlines of Psychology*. Leipzig: W. Engelmann.
Yablo, Stephen (1990): "The real distinction between mind and body", *Canadian Journal of Philosophy* 16, pp. 149–201.

Name Index

Allen, Grant 109
Allen-Hermanson, Sean 74
Ancell, Aaron 166
Anjum, Rani Lill 191
Anscombe, Gertrude E.M. 195
Aristotle 78
Armstrong, David 82, 90, 91, 101, 107
Aydede, Murat 175

Baars, Bernard J. 2, 17, 162
Barrett, Lisa (Feldman) 174
Baumgartner, Michael 197
Bayne, Tim 104
Beebee, Helen 191
Bennett, Karen 62
Berkeley, George 54
Bickle, John 45
Bisiach, Edoardo 163
Bliss-Moreau, Eliza 174
Block, Ned 58, 90, 91, 92, 156, 158, 159, 162, 164, 214
Bogen, Joseph E. 17
Boyd, Richard 61
Bornstein, Robert F. 161
Brandom, Robert 41
Brentano, Franz 41
Bringsjord, Selmer 77
Brooks, Rodney 113
Burge, Tyler 69
Byrne, Alex 91

Cacioppo, John T. 174
Campion, John 161
Callender, Craig 191
Carroll, James M. 174
Carruthers, Peter 3, 55, 58, 80, 82, 87, 91, 107, 111, 213
Chalmers, David 2, 17, 18, 19, 20, 21, 25, 26, 27, 40, 58, 59, 74, 75, 213
Charland, Louis 173
Churchland, Patricia 55
Cohen, Jonathan 191
Collins, John 197
Colombetti, Giovanna 173

Coren, Stanley 109
Crick, Francis 2, 17, 18, 94

Damasio, Antonio 104, 173
Darwin, Charles R. 137, 138, 139, 140, 143
Davidson, Donald 28, 40, 60
Davies, Martin 160
Dawkins, Richard 143
Dehaene, Stanislas 5, 6, 7, 132, 166
DeLancey, Craig 189
Dennett, Daniel C. 2, 26, 32, 40, 55, 58, 77, 91, 174, 175, 213
Descartes, René 1, 43, 58, 59, 70, 82, 220
Diener, Edward 174
Dowe, Phil 198
Dretske, Fred 83, 90, 91, 107, 157, 158, 159, 160, 161, 162, 163, 212, 213, 214

Eccles, John 60
Elisabeth, Princess of Bohemia 59, 220
Esfeld, Michael 10, 28, 41, 42, 43, 68, 191, 207

Feigl, Herbert 56
Feinberg, Todd 30, 31, 34, 126
Feyerabend, Paul 55
Fisher, Ronald A. 200
Flanagan, Owen 77, 78, 162
Flohr, Hans 17
Fodor, Jerry 57, 61
Foster, John 59
Freud, Sigmund 94

Gardner, Howard 2
Gassendi, Pierre 56
Gennaro, Rocco 82, 91
Giacometti, Alberto 171
Gillett, Carl 42
Goodale, Melvyn A. 17, 18
Gould, Stephen J. 146, 148, 149, 150
Grantham, Todd 153
Güzeldere, Güven 94, 95

Name Index

Harnad, Steven 77
Hasker, William 60
Heidegger, Martin 1
Hellman, Geoffrey P. 57
Hill, Christopher S. 56
Hirnstein, William 162
Hitchcock, Christopher 191
Hobbes, Thomas 45, 56
Hobson, John Allan 17
Hufendiek, Rebekka 173, 174
Hume, David 1, 191, 192, 193, 194, 195, 196, 197
Humphrey, Nicholas 162
Humphreys, Glyn W. 160
Husserl, Edmund 1
Huxley, Thomas 24
Hyslop, Alec 135

Jackson, Frank 48, 60, 63, 105
James, William 2, 94, 173
Jennett, Bryan 99

Kant, Immanuel 1, 194
Kim, Jaegwon 41, 50, 52, 60, 61, 62, 63, 203, 217
Kimura, Motoo 138
Kinsbourne, Marcel 55
Kirk, Robert 33
Kistler, Max 198
Koch, Christof 17, 18, 26, 59, 94
Kriegel, Uriah 107
Kripke, Saul 48

Lamarck, Jean-Baptiste 138
Larsen, Randy J. 174
Larsen, Jeff T. 174
Laureys, Steven 99
Leibniz, Gottfried W. 1, 56, 60
Lewis, David K. 57, 192, 200, 217
Lewontin, Richard 139, 146, 150
Levine, Joseph 57, 58, 91, 107, 213
Lindquist, Kristen A. 174
Llinás, Rodolfo R. 17
Locke, John 1
Loewer, Barry 191, 203
Logothetis, Nikos K. 17

Lowe, Jonathan 69
Lycan, William 2, 55, 90, 91, 107, 213

Mallatt, Jon 30, 31, 34, 126
Mancuso, Stefano 73
Marcel, Anthony J. 162, 163
McGinn, Colin 8, 26, 60, 161
McLaughlin, Brian P. 56, 62
Mellor, David 52, 195
Menzies, Peter 191
Merker, Bjorn 104
Merleau-Ponty, Maurice 1
Metzinger, Thomas 17, 25
Mill, James 1
Mill, John Stuart 1, 200
Mills, Eugene 52
Milner, A. David 17, 18, 161
Mumford, Stephen 191

Nagel, Thomas 13, 48, 59, 74, 106, 109, 213
Neander, Karen 81, 91, 145
Neisser, Ulric 2
Nelkin, Norton 90, 91
Newman, James 17
Newton, Isaac 42, 68
Nichols, Shaun 153
Noel, Ron 77
Nolan, Christopher 117
Norris, Catherine J. 174

Papineau, David 42, 56, 67, 91
Pappas, George 55
Penfield, Wilder 163
Penrose, Roger 2
Pinker, Steven 189
Pittman, Thane S. 161
Place, Ullin T. 56
Pöppel, Ernst 161
Polger, Thomas 77, 78
Popper, Karl 60
Prinz, Jesse 173
Putnam, Hilary 57, 61

Ramachandran, Vilayanur 162
Ratcliffe, Matthew 173

Royo, Luis 170, 171
Rorty, Richard 55
Rosenberg, Gregg 74
Rosenthal, David 80, 81, 82, 83, 87, 91, 103, 107, 213, 214
Rugg, Michael D. 161
Russell, Bertrand 54, 192
Russell, James A. 173, 174
Russo, Andrew 203
Ryle, Gilbert 59

Seager, William 74
Searle, John R. 5, 6, 9, 11, 12, 14, 36, 47, 48, 49, 50, 60, 92, 93, 95, 116, 143, 159, 163, 164
Sheinberg, David L. 17
Shoemaker, Sydney 58, 91
Siewert, Charles 213
Skinner, B. Frederic 2
Smart, John J.C. 56
Sober, Elliott 137, 138, 139, 141
Sosa, Ernest 191
Spinoza, Baruch 194
Stich, Steven 158
Stout, George 94
Strawson, Galen 3, 54
Suárez, Mauricio 195
Sutherland, Stuart 95
Swinburne, Richard 59

Teasdale, Graham 99
Tellegen, Auke 174
Thompson, Frank W. 57
Thorndike, Edward 184, 185, 186, 187
Titchener, Edward 2, 94
Tononi, Giulio 59
Tooley, Michael 191
Trumbull Ladd, George 93, 95
Turing, Alan 166
Tye, Michael 90, 107, 213

Van Gulick, Robert 1, 2, 60, 91, 162, 163
Velmans, Max 32
Viola, Alessandra 73
von Helmholtz, Hermann 2
Vrba, Elisabeth 146, 148, 149

Wachowski, Andy & Lana 117
Watson, John 2
Watson, David 174
Weber, Marcel 141, 142
Weiskrantz, Lawrence 161
Wheeler, Michael 112
Wilberg, Jonah 91
Wild, Markus 40, 42, 68, 184, 207
Wilkes, Kathleen V. 55
Wundt, Wilhelm 2, 174

Yablo, Stephen 59

www.ingramcontent.com/pod-product-compliance
Lightning Source LLC
Chambersburg PA
CBHW070315240426
43661CB00057B/2645